MEMS Lorentz Force Magnetometers

Cesare Buffa

MEMS Lorentz Force Magnetometers

From Specifications to Product

Springer

Cesare Buffa
Dipartimento di Elettronica
Informazione e Bioingegneria
Politecnico di Milano
Milano, Italy

ISBN 978-3-319-59411-8 ISBN 978-3-319-59412-5 (eBook)
DOI 10.1007/978-3-319-59412-5

Library of Congress Control Number: 2017941507

Printed on acid-free paper

This Springer imprint is published by Springer Nature
The registered company is Springer International Publishing AG
The registered company address is: Gewerbestrasse 11, 6330 Cham, Switzerland

This book is dedicated to my sister Beatrice and to my parents Manuela and Rocco.

Preface

Nowadays sensors are part of everyday life in a wide variety of fields: scientific applications, medical instrumentation, industrial field, ... and, last but not least, popular mass production and low-cost goods, like smartphones and other mobile devices. Markets and business behind the field of sensors are quite impressive. A common trend for consumer applications is miniaturization which requires, on one side, a lot of research, development efforts, and resources but, on the other hand, allows costs and final application size reduction. In this scenario scientific community and industries are very active to drive innovation.

I started my research activity in 2009 in the field of imaging sensors and microelectromechanical sensors. During this period I had several opportunities to write technical papers and articles, to participate in conferences, to exchange knowledge with technical experts, to advise students, and to give some lectures at the university. The fact that I particularly enjoyed sharing my knowledge made me think of writing a book one day in my life.

This book deals with compasses for consumer applications realized in micro-technologies, describing a full path from specifications to prototype and preparing the way to industrialization and commercialization. This book does not pretend to be fully comprehensive in each single step of the development phase, rather it is meant to show the overall process, all the way from some specifications (not much more than a few numbers) to a concrete working sensing system. The book is intended for a wide audience, especially for researchers who want to develop their prototypes and people who want to go beyond academic research toward industrialization. The book also includes some introductory concepts about microelectromechanical systems and therefore can be considered for undergraduate and graduate courses too. The author wishes that this manuscript fulfills his target. Even though some sections deal with theoretical aspects, the book wants to be primarily a "guide" (independent from the topic itself of magnetic field sensors) conveying a practical approach, with pragmatic guidelines and design choices, and keeping an eye on the final target, industrialization and mass production. This book cannot describe every single detail (this process involves a lot of people with different skills and expertise according to the development phase) but it is intended to mention all the major steps of this research and development process, highlighting where possible critical aspects or weak points can arise and how to forecast and handle them. I would like to go through this, taking advantage of my years spent in the scientific world as a researcher but also targeting a development for mass production products. These two worlds are at the same time complementary and in contrast. This book deals with Lorentz force sensors implemented in MEMS technology but general guidelines, methodology, and development flow are applicable to any other development project related to microsensors.

This book is organized into eleven chapters.

Chapter 1 introduces the topic of the book, highlighting motivations and objectives.

Chapter 2 shows a methodology from specifications to the definition of a concept. Main requirements for consumer market products and specifications are analyzed in detail, followed by the choice of transduction principle. Finally, a brief overview of the state of the art is proposed, to review some scientific works published in the literature. Main features of each work are summarized in a table and compared to given specifications for the development of the prototype proposed in this book.

Chapter 3 deals with Lorentz force magnetometers and their working principle. After an introduction about the dynamic of MEMS and capacitive readout principle, basic properties of Lorentz force transduction principle are presented, applied to a micromachined suspended structure. A new design approach of mechanical sensors at a fixed resonance frequency is shown, supported by the study and development of a theory about sensitivity in the free molecular flow regime, which applies for the pressure range of typical industrial packages. Finally, a brief introduction to thermomechanical noise is provided in order to discuss intrinsic resolution of devices.

In Chap. 4 a behavioral model at system level is illustrated, which is developed to analyze magnetometers dynamics and to set specifications for major system subblocks. Then noise budget partitioning and power budget partitioning are discussed.

Chapter 5 is about mechanical devices design, based on specifications derived at system level and supported by finite element simulations. After an introduction to a micromachining fabrication process, devices layouts and photographs are described.

Chapter 6 deals with the design of an instrument for electromechanical characterization. A detailed description of this measurement platform is provided together with its specifications, design, and performance.

In Chap. 7 electromechanical measures—both static and dynamic—are reported in order to validate the proposed theory about sensitivity and to characterize mechanical properties of devices.

Chapter 8 deals with the development of a setup for magnetic field measurements using implemented devices. After identifying main specifications for driving sensors in order to avoid unwanted additional electrostatic forces, the first part of this chapter shows the design of a bipolar differential current generator. The following sections focus on the readout of the capacitance variation analyzing different solutions in terms of complexity of the architecture and noise and taking into account the overall assembling of mechanical part, driving and readout electronics. Finally experimental measures are reported.

Chapter 9 reports about ASIC architectures and design. A Verilog-A model is also shown to support electromechanical coupled simulation in Cadence environment. Experimental measurements with a multi-chip (MEMS + ASIC) prototype are reported.

Chapter 10 gives an overview of main steps toward industrialization of the presented prototype.

Finally conclusions are reported in Chap. 11.

The work presented in this book is the result of scientific research during years 2010–2012 when the author was a Ph.D. student at Politecnico di Milano in Italy, in cooperation with STMicroelectronics and visiting researcher at Berkeley Sensor & Actuator Center (BSAC), University of California, Davis, CA.

Milano, Italy Cesare Buffa

Acknowledgments

The origin of this book is when I started my Ph.D. program at Politecnico di Milano in Milano, Italy, where I also began my university studies in 2004 and developed my B.Sc. and M.Sc. theses. Overall I spent a learning-full, productive, and fruitful period of about eight years and therefore I would like to start thanking a lot of people from this environment, to whom I am very grateful.

My first thanks are due to my advisor, professor Antonio F. Longoni, for the opportunity to join his team, to work in the laboratory—SanDLab—and to be involved in many research topics, all the way from the beginning of my M.Sc. thesis to my Ph.D. dissertation. Professor Longoni's personal help and technical teaching supported me for 4 years and he was able to hand his fascinating passion to examine in depth all the experimental evidence to me. And I am glad of his current pleasant friendship.

My special thanks to my co-advisor and mentor, professor Giacomo Langfelder, who has been the essence of technical advancements and who has always been able to motivate people, showing his admirable devotion to research.

I am particularly thankful to professor David A. Horsley for the opportunity to join Berkeley Sensor & Actuator Center (BSAC), University of California, Davis, CA, as a visiting researcher from January 2012 to July 2012; and to MEMSLab people for their warm welcome and friendship.

I also would like to thank professors Giuseppe Bertuccio and Federico Zaraga for research and teaching collaborations during these years.

I am grateful to professors Attilio Frangi, Alberto Corigliano, and Claudia Comi of the Department of Civil and Environmental Engineering for their fruitful cooperation.

Many thanks are also due to Dr. Anu Kärkkäinen, VTT Technical Research Centre of Finland, for her careful review and for precious technical advices to strengthen my Ph.D. thesis.

Secondly, I thank Dr. Ernesto Lasalandra—STMicroelectronics, Cornaredo, Italy—for his technical guidance focused on industry and for access to facilities.

Thirdly, I want to thank all my colleagues at SanDLab laboratory at Politecnico di Milano for the relaxing, friendly, and encouraging atmosphere and all the people

I have had the pleasure to work with. A special "thank you" to people who helped me during the design, fabrication, assembly, and measurements; among them, my followers Stefano Dellea and Alessandro Caspani. Dr. Caspani deserves special thanks and gratitude for the time he dedicated to review this book and for his unique friendship. Additionally, Sergio Masci for wire bondings and PCB fabrication; my M.Sc. thesis students Marco Rossi, Roberta Mazzola, Alice Pelamatti, Nicola Aresi, and Daniele Laudi; Ernestina Parente and Emilia Aloisio for administrative procedures; and Vasco Previtali and Roberto Resmini for IT services.

I thank my friends with whom I spent my university times: Nicole De Togni, Alessandro Tocchio, Alberto Recalenda, Giulia De Donno, Giorgio Carlo Brambilla, and Stefania Butera—to share disappointments and gratifications. Special thanks to Simonetta Grandi, and Alessandra Migliau with her husband Alberto, for their long-standing and strong friendship. Many thanks to new friendships, born after university times, with Andrea Lavarda, Stefano De Filippis, Marco De Din, Francesco Solazzi, Giulio Fragiacomo, Paolo Toniutti, Nicola Serra, Marco Bresciani, Giuseppe De Falco, and Roland Felderer.

Last but not least, my warmest thanks to my family whom this book is dedicated to. I would like to thank my parents for funding my studies and education and my sister to be the best sister!

Milano, Italy Cesare Buffa

Contents

Acronyms

Lists of Abbreviations and Symbols

AMR	Anisotropic Magneto-Resistance
CAGR	Compound Annual Growth Rate
CMOS	Complementary Metal-Oxide Semiconductor
DAC	Digital to Analog Converter
DAQ	Data Acquisition Board
DDS	Direct Digital Synthesizer
DOF	Degree of Freedom
DRIE	Deep Reactive Ion Etching
DRM	Design Rule Manual
EMC	Electromagnetic Compatibility
EPL	Epitaxial Polysilicon Layer
FSR	Full Scale Range
FEM	Finite Element Models
GPIB	General Purpose Interface Bus
IC	Integrated Circuits
IMU	Inertial Measurement Unit
LDO	Low Dropout Regulator
LPF	Low Pass Filter
MCP	Mechanical Characterization Platform
MEMS	Microelectromechanical Systems
NEMS	Nanoelectromechanical Systems
PCB	Printed Circuit Board
PECVD	Plasma Enhanced Chemical Vapor Deposition
SoC	System-on-Chip
SNR	Signal to Noise Ratio
ThELMA	Thick Epitaxial Layer for Micro-gyroscopes and Accelerometers
TIA	Transimpedance Amplifier
TPL	Thin Polysilicon Layer
VLSI	Very Large Scale Integration

Chapter 1
Introduction

1.1 Motivation and Context for Research

Microelectromechanical systems (MEMS) originally referred to microscale systems made of both electrical and mechanical moving parts, being able to perform something interesting or useful. They are μ-sized 3D structures, realized with lithographic processes, which can interact with external world and they are typically integrated with electronic circuits. The presence of mechanical moving parts let them be sensors and actuators of our environment; they are reliable and cost effective.

In 1959 the Nobel Laureate Richard Feynman gave his inspiring talk "There's Plenty of Room at the Bottom," inviting the scientific community to exploit the huge amount of opportunities offered by the unexplored field of micro- and nano-fabrication and emphasizing the possibility of manipulating matter on atomic scale. This talk may be considered to some extent the birth or, at least, the precursor of MEMS world and evolution [1]. Around the end of the 1970s first micromachined devices were developed by different companies, like Hewlett-Packard (HP), Analog Devices (ADI), and Texas Instruments (TI). HP was the first to develop a proprietary technology to manufacture inkjet cartridge, TI developed digital micromirrors for light manipulation, and ADI is commonly recognized as the pioneer for MEMS accelerometers. European companies like Robert Bosch GmbH and STMicroelectronics played an important role starting at the beginning of the 1990s—Bosch process for Deep Reactive Ion Etching (DRIE) was patented in 1994—and they are still important players on the global market for different types of sensors.

Nowadays, MEMS are an expanding technology involving many different markets such as automotive (e.g., tires pressure monitoring systems, airbag systems, ESP, self-driving capability, etc.), consumer (e.g., gaming, navigation, image stabilization, augmented reality, etc.), home automation (e.g., distance speech interaction, powering wireless sensors where solar energy is not available, etc.),

C. Buffa, *MEMS Lorentz Force Magnetometers*,
DOI 10.1007/978-3-319-59412-5_1

healthcare (drug delivery, cardiac monitoring, implantable neurostimulation, air quality monitoring, etc.), and industrial (building stability, structural monitoring, etc.).

While, up to few years ago, MEMS were thought as single sensor or actuator, the present trend sees MEMS to be a combination of functions. Inertial Measurement Units (IMUs) are multi-parameter systems in one single package based on multi-axis MEMS and they play a fundamental role in several fields of applications like navigation units in smart mobile phones and other consumer electronic goods [2]. The realization of an inertial measurement unit is of great interest for civil and military aviation, space satellites, trains, ships, unmanned or remote operated vehicles, stabilization systems, consumer electronics, and several other applications [3, 4]. In particular, as far as consumer applications are concerned, an increasing number of smartphones and mobile devices provides users with location- and orientation-based services in addition to "more traditional" functionalities based on navigation. Navigation is becoming a must-have feature in portable devices and the presence of a compass also makes location-based augmented reality emerge: a street map or a camera image could be overlaid with highly detailed information about what is in front of the user. To make these features possible both industries and scientific research are focusing on three-axes magnetic field sensors (also called *magnetometers*). And, like the rest of electronics industry, compasses are required to scale down, get cheaper, and more energy efficient [5, 6].

In this scenario, the integration of a three-axis accelerometer, a three-axis gyroscope, a three-axis magnetometer, and a pressure sensor, all based on the same MEMS process, can result in a "10-degrees of freedom (DOF)" high resolution, low-cost, and low-power miniaturized system for position, motion, heading, and altitude monitoring, representing a challenge for research. According to recent market reports, fusion of sensors is starting to be widely used in consumer fields and inertial MEMS market is booming and expected to register a CAGR growth of about 7.8% by 2020 [7].

The state of the art for IMUs integrated using MEMS processes only is represented by a 6-axis MEMS unit for acceleration and angular rate sensing [8, 9]. Dominant technologies used to integrate magnetic sensing elements in IMUs which are currently on the market[1] are not based on micromachining processes yet. The most common implementations of electronic compasses are based on Hall effect and on anisotropic magneto-resistance (AMR) technologies. For instance, the multi-axes systems for motion and magnetic field sensing described in [10, 11] include accelerometers or gyroscopes based on MEMS technology but magnetometers based on AMR [12].

Thus, AMR and Hall effect technologies still represent the most common implementations of electronic compasses. Devices based on AMR technology require the deposition of specialized magnetic materials into standard industrial processes, they suffer from magnetic hysteresis, require to be continually degaussed

[1] At the time this research activity began.

and, finally, they are sensitive only to in-plane fields. The magnetometer for the Z-axis is the same as for X and Y axis but in order to integrate a three-axis AMR-based sensing system into an IMU, a vertical assembling for the Z-axis device is required, which has a strong impact on the packaging cost and it represents an obstacle for ultra-thin packages. Large volume is then wasted in height due to the required out-of-plane assembling, reducing the integrability of the IMU in ultra-thin plastic packages. Even though Silicon Hall effect sensors avoid introduced problems arising from the integration of magnetic materials, their power consumption and resolution performance are relatively limited with respect to AMR devices.

The research on an alternative approach, Lorentz force-based MEMS magnetometers (in particular for the Z-axis element), is encouraged by the possibility to integrate magnetic sensors in the same MEMS technology already used for gyroscopes and accelerometers, so to design a complete 9-axis IMU in a single standard process. Moreover, among advantages of the integration of compasses in MEMS technology are sufficiently high resolution, sufficiently low power consumption, and no need to use ferromagnetic materials. In a very interesting intermediate step, a 9-DOF IMU can be realized with a 2-axis AMR magnetometer and a 7-DOF MEMS die, with a consequent reduction of packaging costs, open the path to vertical scalability and improve mechanical reliability.

Several solutions to combine MEMS and integrated circuits (IC) have been proposed in the literature and two major approaches can be identified: (1) system-on-chip (SoC) solutions, where mechanical elements and electronics are monolithically integrated on the same wafer; (2) multiple chips (multi-chip) solutions, where mechanical elements are manufactured on dedicated dies which are bonded to ICs [13].

1.2 Objectives

The main objective of this work is to experimentally demonstrate the possibility to integrate a magnetic field sensing system using a multi-chips solution based on standard industrial processes and to achieve sensing specifications, which are required for navigation in consumer products. The choice of multi-chip approach is driven by its higher flexibility and lower development costs with a quite rapid system development cycle.

1.3 Research Contribution

The research path followed during this project consisted of study and development of a magnetic field sensing system for consumer applications starting from specifications. First, the work focused on system level modelling and on the design of MEMS magnetometers trying to provide design criteria for consumer products. Basing on the available technology, a novel design approach for MEMS magnetometers insensitive to ambient acoustic noise is investigated and characterized, showing that the specifications required for consumer applications can be fulfilled with a very compact device by using standard industrial processes. Second, research activity focused on the design of both driving and readout electronics keeping an eye at a system level to guarantee resolution, power consumption, linear range required by the application. The resulting prototype is one of the first to show the feasibility of a Z-axis Lorentz force-based magnetometer implemented with a double chips solution and using standard processes: one chip hosts the mechanical element and in the second chip readout electronics is integrated.

References

1. R.P. Feynman, Eng. Sci. **23**(5), 23 (1960)
2. R. Dean, A. Luque, IEEE Trans. Ind. Electron. **56**(4), 913 (2009). doi:10.1109/TIE.2009.2013691
3. L. Robin, M. Perlmutter, IMU and high performance inertial MEMS 2011. Technical Report, Yole Développement (2011). www.yole.fr
4. S.H. Won, F. Golnaraghi, W. Melek, IEEE Trans. Ind. Electron. **56**(5), 1782 (2009). doi:10.1109/TIE.2008.2010166
5. C. Brigante, N. Abbate, A. Basile, A. Faulisi, S. Sessa, IEEE Trans. Ind. Electron. **58**(8), 3234 (2011). doi:10.1109/TIE.2011.2148671
6. R. Antonello, I. Nogarole, R. Oboe, in *2011 IEEE International Symposium on Industrial Electronics (ISIE)* (2011), pp. 2189–2194. doi:10.1109/ISIE.2011.5984500
7. Inertial Measurement Unit (IMU) Market by Component - Accelerometer, Gyroscope, Magnetometer, by Application - Commercial, Defense, Industrial & Marine, by Gyro Technology-RLG, FOG, MEMS, HRG, Vibrating Gyro, by Platform – Airborne, Land, Naval, and by Geography - Forecasts & Analysis to 2014 - 2020. Technical Report, MarketsAndMarkets (2015). http://www.marketsandmarkets.com/Market-Reports/inertial-measurement-unit-market-148851976.html
8. STMicroelectronics, Technical Datasheet (2011). http://www.st.com
9. InvenSense, Technical Datasheet (2012). http://invensense.com/mems/gyro/documents/PS-MPU-6100A.pdf
10. STMicroelectronics, Technical Datasheet (2009). http://www.st.com
11. InvenSense, Technical Datasheet (2012). http://invensense.com/mems/gyro/mpu9150.html
12. M. Afzal, V. Renaudin, G. Lachapelle, in *2011 International Conference on Indoor Positioning and Indoor Navigation (IPIN)* (2011), pp. 1–10. doi:10.1109/IPIN.2011.6071947
13. A.C. Fischer, F. Forsberg, M. Lapisa, S.J. Bleiker, G. Stemme, N. Roxhed, F. Niklaus, Microsyst. Nanoeng. **1**, 15005 (2015). doi:10.1038/micronano.2015.5

Chapter 2
From Specifications to Concept

This book is organized in a way that illustrates the full path of development of a real product, which, in its early stage, implies scientific research and the development of a prototype, starting from some requirements. Specifications are typically given by a customer or derived according to the target application. Discussions and agreement on specifications can represent an iterative process which develops in different ways for each case. Even though some specifications can be negotiated and refined, some other parameters, like technology, costs, and time-to-market, are particularly stringent in consumer markets projects. This chapter introduces how to 'translate' product specifications into the definition of a concept with its blocks specifications and proposes a possible flow and methodology for the development of a magnetic field sensor. The proposed case is based on the development of a Z-axis Lorentz force-based MEMS magnetometer but the methodology is meant to be applicable to the development of other types of sensors.

2.1 Development Steps and Methodology

In projects with a relatively short-time development, i.e., projects for consumer market, it is important to have a solid and wide overview of all possible implementations and to target the most suitable one from the beginning, minimizing the risk of changing approach and design strategies half way in the process. Indeed this would cause a delay in the development of the project which is typically not compatible with the time-to-market of consumer products. According to author's experience, as far as the phase "from specifications to concept" is concerned, main steps for a successful and solid development are here listed and they will be discussed in chapters of this book:

C. Buffa, *MEMS Lorentz Force Magnetometers*,
DOI 10.1007/978-3-319-59412-5_2

1. Analysis of requirements and specifications.
2. Study of suitable transduction principles.
3. Research about the state of the art.
4. Study and development of theoretical background (if needed).
5. Identification of a feasible concept.
6. System level design based on the chosen concept.
7. Translation of product specifications into block specifications.

System level design is discussed in Chap. 4; subsequently in Chap. 5 the following steps about device design will be shown.

2.2 Specifications for Consumer Applications

Electronic compasses integrated in mobile devices are used for orientation, positioning, and navigation applications. Thus, the measurement of earth magnetic field components along three directions is required to precisely determine magnetic field vector in device operating position. Three axes are important to make the portable device orientation-independent. If the magnetic field lays in the plane of the handheld device, then the compass heading could be computed from the inverse tangent of the ratio of the two horizontal magnetic field components. The third axis avoids that users have to keep the device at a leveled position. Moreover, accelerometers, which are usually integrated in the same IMU with magnetometers, are used to implement a tilt-compensated compass [1, 2].

Main requirements for mentioned portable compasses are

1. Adequate sensitivity and resolution to measure earth magnetic field. Figure 2.1 reports earth field total intensity.
2. Linear full scale range: even though earth maximum magnetic field magnitude can reach about 100 μT, it is important to have a much higher full scale range to cope with local high magnetic field. For example, in presence of vehicles or other ferromagnetic sources in the surroundings of the mobile device, a stronger magnetic field should not saturate the compass to allow a proper detection of earth magnetic field.
3. Low power consumption, a crucial parameter in battery-hungry portable devices.
4. Small package size for the integration in portable devices. The trend is to integrate a compass with other inertial sensors in one single package.
5. Low operating voltage; a typical operating voltage for standard CMOS circuits available in portable devices is 1.6–3.6 V.
6. Compatibility with standard fabrication processes to optimize mass production costs.
7. Audio bandwidth EMC: immunity to disturbers in audio bandwidth to avoid coupling mechanisms and crosstalk from microphones, speakers, and MP3 players. Immunity must be ensured not only for electrical aggressors but also to prevent coupling to propagated mechanical waves.

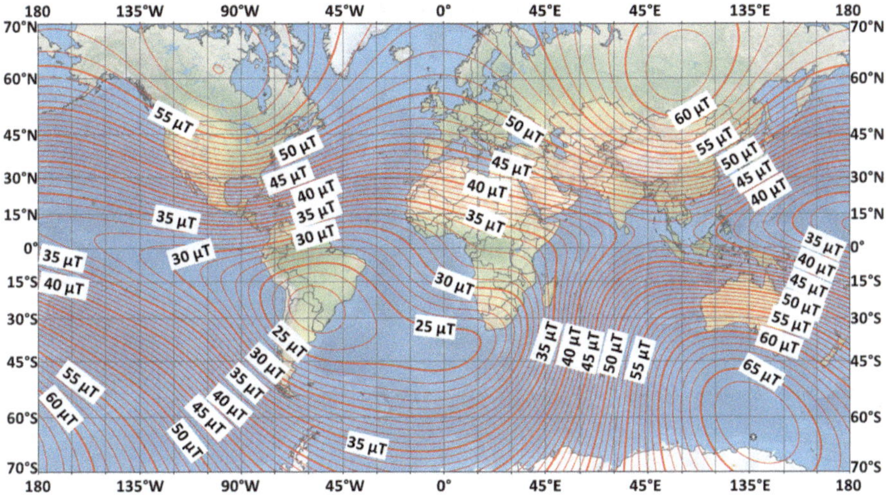

Fig. 2.1 Mercator projection chart showing main field total intensity. Contour interval: 1000 nT. Map developed by NOAA/NGDC & CIRES; http://ngdc.noaa.gov/geomag/WMM; map reviewed by NGA and BGS. Published December 2014. Courtesy of NOAA National Centers for Environmental Information

Table 2.1 Main specifications for a prototype of a *Z*-axis magnetometer for consumer applications

Magnetometer specifications	
Voltage supply	1.6–3.6 V
Total current consumption	300–400 µA
Bandwidth	< 50 Hz
Full scale linear range	±100 µT
Resolution	< 1 µT (~2°/3°)
Resonance frequency	$f_r \geq 20$ kHz
Pressure	Standard industrial packaging
Architecture	Multi-chip
MEMS active area	<1 mm × 2.5 mm
ASIC size	<2 mm^2

8. Operating bandwidth according to applications: typically in the range of a few tens of Hz.
9. Calibration to compensate for production spread.

Specifications are reported in Table 2.1.

2.3 Choice of Transduction Principle

The development of this prototype is based on Lorentz force transduction principle because of its main advantage, compared to other transduction principles, to detect magnetic field: indeed no specialized magnetic materials are required, being a key factor for an implementation with any standard micromachining technology. Additionally, compared to Hall effect-based sensors, which are commercially available and used in smartphones, they do not suffer from magnetic hysteresis and do not need flux-concentrator.

2.3.1 Lorentz Force Transduction Principle

Several works on MEMS magnetometers based on Lorentz force transduction principle are presented in scientific literature (see Sect. 2.4). In all these devices, in presence of a magnetic field B, a driving current I, flowing in a suspended structure orthogonally to the direction of B, determines a force in a direction orthogonal to the plane of both B and I: $\mathbf{F}_\mathrm{L} = i\mathbf{l} \wedge \mathbf{B}$. The magnitude of the resulting force on a suspended mass is given by:

$$F_L(t) = I(t) \cdot l \cdot B(t) \,, \tag{2.1}$$

l being the effective length where the Lorentz force acts. Figure 2.2 shows a clamped–clamped beam, the simplest suspended element to have a flow of current. If a current flows through this suspended structure and an external magnetic field is present, this beam moves according to arrows reported in Fig. 2.2.

2.3.2 Position Sensing Techniques

Though the induced motion can be sensed in different ways (e.g., through the change of resistance in piezoresistors [3, 4], using optical techniques based on microlasers

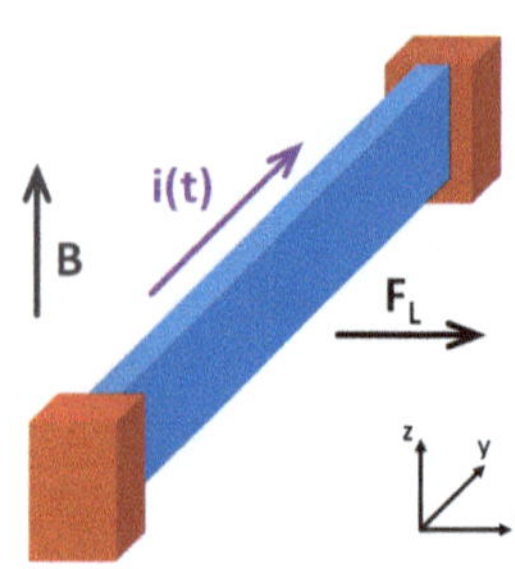

Fig. 2.2 Schematic view of the Lorentz force principle acting on a suspended clamped–clamped beam

[5, 6] or through the change in the resonance frequency of suitably designed resonators [7]), a very popular position interface is represented by capacitive readout [8]. Piezoresistive sensing allows a very high level of miniaturization with good sensitivity but they suffer from noise of piezoresistors and, at least, a dedicated lithography step is needed. Optical sensing is good for out-of-plane measurements, allows measurements with high resolution but it requires the integration of an optical source (typically a laser), a photodetector, and the deposition of some dedicated materials. Therefore this readout technique results in a more expensive and complex system. On the other hand, capacitive readout is easy to be integrated with a standard process because a basic cell can be built simply by facing two structures of polysilicon, either in a vertical or a horizontal orientation. Among other advantages of capacitive readout are its stability over temperature and process corners and no intrinsic sources of noise. Yet, the main drawback is given by electrostatic forces and their dependencies on biasing conditions; more details are given in Sect. 3.1.2.

Additional details about magnetometers working principle and capacitive readout focusing on the specific device architecture developed during this project are reported in Sect. 3.2.

2.4 State of the Art

Many reports and reviews about magnetic sensors and their applications are available in the scientific literature [9, 10]: the purpose of these papers is mainly to provide readers with an introduction to all existing and more common magnetic field transduction techniques together with some technological aspects and applications they are intended for. A more specific overview of resonant magnetic field sensors based on MEMS technology is given by Herrera-May et al. [11].

In this section some recent[1] and/or significant scientific works are reported focusing on papers dealing with Z-axis magnetometers. The purpose of this brief scientific literary review is to identify some solutions which are already available at the time when the development of a new prototype begins. Both mechanical element and circuitry are considered and it is a good practice to compare their performance with given specifications. After a short introduction to each work, highlighting its main characteristics, a table sums up more important parameters to consider as a reference for this project. Works are introduced following chronological order of publication.

A first example of a Lorentz force magnetometer based on MEMS technology with capacitive sensing was presented by Emmerich and Schöfthaler [12], Emmerich et al. [13] of the Sensor Technology Center (STZ) of Robert Bosch GmbH, Reutlingen, Germany in 1999. The proposed mechanical element, operating at resonance to exploit Q-amplification, consists of a central beam with parallel-

[1] At the time research activity behind this book began.

plates electrodes, suspended by four folded spring. A thin layer of aluminum is deposited on top of moving structure to reduce electrical resistance and to provide a constant voltage across all movable fingers minimizing attractive forces which arise from electrical potential difference between fixed and movable electrodes. The device has an area of 1300 μm × 500 μm with 32 electrode fingers and it is implemented using Bosch standard surface micromachining [14] and packaged with a nominal pressure of 1 mbar. Experimental results are provided using discrete components electronics for capacitance to voltage conversion and device driving. There are not any specific details about power consumption of the overall system (even though current flowing in mechanical element is known).

This work represents a significant reference since the used technology is very similar to one available for the development of prototypes shown in this work (details are given in Sect. 5.2.1); Bosch technology has the main advantage of metal deposition on top of released structures. Resonance frequency of this prototype, about 1.3 kHz, is definitely lower than specifications given for this project.

A second prototype of magnetometer was published in [15] by Izham et al. of University of Birmingham in 2003. Even though resonance frequency of about 9 kHz is not so lower than the specification for this project, intrinsic current consumption of 29 mA per spring is by far out of specifications. Moreover, the paper does not give any details about electronics and so no overall system considerations are provided in this paper.

A third example concerning both out-of-plane and in-plane Lorentz force magnetometers was proposed by Kyynäräinen et al. [16] of the Technical Research Centre of Finland, in 2008. They integrate micromechanical sensing elements for all three axes in one silicon chip using a process which allows metal deposition on top of moving structures. The specific element proposed for out-of-plane field detection is based on a double-ended tuning fork architecture, working at resonance with high quality factor (about 10,000 for pressure in the range 10–100 μbar). The peculiarity for this implementation is the use of multi-turn excitation coils: on top of one single released element many metal paths (either 5 or 10 for *Z*-axis sensitive element) for current are deposited enhancing the sensitivity without increasing power dissipation. Resonance frequencies for these devices are around 50 kHz, satisfying the requirement to work out of the audio bandwidth. On the other hand, such a high *Q*-factor limits mechanical bandwidth to about 5 Hz with a consequent longer settling time when the device is switched on.

A fourth paradigmatic example of a parallel-plates magnetometer was proposed by Thompson and Horsley [17] working at Mechanical and Aerospace Engineering at University of California, Davis in 2009. Their first prototype, fabricated using SOIMUMPS foundry process (MEMSCAP Inc.), operates at resonance frequency (f_r = 8468 Hz) at ambient pressure. The architecture of the device is close to the first introduced and folded springs are chosen to allow for longitudinal expansion without stress accumulation. The main drawback of using folded flexures results in reduction of effective Lorentz force. A larger occupied area of about 2200 μm × 700 μm could also be explained because the used process does not have an interconnect layer to route differential capacitors cells.

Table 2.2 List of more recent and/or significant scientific works about *Z*-axis magnetometers

Z-axis Lorentz force magnetometers								
Work	Area (μm × μm)	I (mA)	Noise ($\mu T/\sqrt{Hz}$)	BW (Hz)	Mass (μg)	Q (–)	f_r (kHz)	k (N/m)
Emmerich and Schöfthaler [12]	1300 × 500	0.93	0.20	10[a]	3.6	37	1.3	0.24[a]
Kyynäräinen et al. [16]	2000 × 400	0.1	0.07	2[a]	12[b]	10,000[c]	50	250[b]
Thompson and Horsley [17]	2200 × 700	2.6	1.01	87[a]	6.8	48.8	8.5	19.5
Li et al. [21]	1000 × 200[d]	0.4	0.14	7	0.82[a]	1400	20.55	13.7

For each work many parameters are reported, particularly the noise density performance with respect to the active area of the sensing element, the dissipated current, and the maximum signal bandwidth

[a]Deduced from other parameters given in the paper

[b]Kindly provided by the authors

[c]At a pressure ~ 0.01 mbar

[d]Full size for a 2D sensing element

The same authors also proposed an external solution to boost quality factor, known as *parametric amplification*. It consists of a modulation of the elastic stiffness at twice the natural frequency [18]. As a consequence, there is an increase of the oscillation amplitude at the device resonance frequency and thus an increase of Q. A MEMS resolution is limited by a mix of electronic and mechanical noise and the reduction of electronics noise can be achieved at the cost of increasing the biasing current of the preamplifier and so the power consumption. When micromechanical sensors performance is limited by electronic noise, parametric amplification can improve the system signal to noise ratio because the equivalent noise force resulting from electronic noise is inversely proportional to the force sensitivity. The main drawback is that it requires additional circuitry to drive the spring stiffness [19].

The same research group presented a second generation of devices in [20] and in [21], whose resonance frequency is pushed to higher frequency outside audio bandwidth.

A summary of main parameters and features of considered papers about Lorentz force magnetometers is reported in Table 2.2.

References

1. STMicroelectronics, *AN3192 Application Note. Using LSM303DLH for a Tilt Compensated Electronic Compass* (2010)
2. T. Ozyagcilar, *AN4248 Application Note. Implementing a Tilt-Compensated eCompass using Accelerometer and Magnetometer Sensors.* Freescale Semiconductor (2012)
3. V. Beroulle, Y. Bertrand, L. Latorre, P. Nouet, Sens. Actuators, A **103**(1–2), 23 (2003). doi:10.1016/S0924-4247(02)00317-5. http://www.sciencedirect.com/science/article/pii/S0924424702003175

4. P. Robert, V. Nguyen, S. Hentz, L. Duraffourg, G. Jourdan, J. Arcamone, S. Harrisson, in *2009 IEEE Sensors* (2009), pp. 963–966. doi:10.1109/ICSENS.2009.5398195
5. F. Keplinger, S. Kvasnica, H. Hauser, R. Grossinger, IEEE Trans. Magn. **39**(5), 3304 (2003). doi:10.1109/TMAG.2003.816753
6. F. Keplinger, S. Kvasnica, A. Jachimowicz, F. Kohl, J. Steurer, H. Hauser, Sens. Actuators, A **110**, 112 (2004). doi:10.1016/j.sna.2003.10.025
7. A. Tocchio, A. Caspani, G. Langfelder, IEEE Sensors J. **12**(6), 1719 (2012). doi:10.1109/JSEN.2011.2177657
8. G. Langfelder, T. Frizzi, A. Longoni, A. Tocchio, D. Manelli, E. Lasalandra, Sens. Actuators, A **167**(2), 374 (2011). doi:10.1016/j.sna.2011.02.003
9. J. Lenz, S. Edelstein, IEEE Sensors J. **6**(3), 631 (2006). doi:10.1109/JSEN.2006.874493
10. D. Robbes, Sens. Actuators, A **129**(1–2), 86 (2006)
11. A.L. Herrera-May, L.A. Aguilera-Cortés, P.J. García-Ramírez, E. Manjarrez, Sensors **9**(10), 7785 (2009)
12. H. Emmerich, M. Schöfthaler, IEEE Trans. Electron Devices **47**(5), 972 (2000). doi:10.1109/16.841228
13. H. Emmerich, M. Schöfthaler, U. Knauss, in *MEMS'99. Twelfth IEEE International Conference on Micro Electro Mechanical Systems, 1999* (1999), pp. 94–99. doi:10.1109/MEMSYS.1999.746759
14. M. Offenberg, F. Larmer, B. Elsner, H. Munzel, W. Riethmuller, in *The 8th International Conference on Solid-State Sensors and Actuators, 1995 and Eurosensors IX. Transducers'95*, vol. 1 (1995), pp. 589–592. doi:10.1109/SENSOR.1995.717293
15. Z. Izham, M. Ward, K. Brusnon, P. Stevens, in *Nanotech 2003*, vol. 1 (2003)
16. J. Kyynäräinen, J. Saarilahti, H. Kattelus, A. Kärkkäinen, T. Meinander, A. Oja, P. Pekko, H. Seppä, M. Suhonen, H. Kuisma, S. Ruotsalainen, M. Tilli, Sens. Actuators, A **142**(2), 561 (2008). doi:10.1016/j.sna.2007.08.025
17. M. Thompson, D. Horsley, in *2009 IEEE Sensors* (2009), pp. 992–995. doi:10.1109/ICSENS.2009.5398216
18. M. Thompson, D. Horsley, in *TRANSDUCERS 2009. International Solid-State Sensors, Actuators and Microsystems Conference, 2009* (2009), pp. 1194–1197
19. L. Grasser, H. Mathias, F. Parrain, X.L. Roux, J. Gilles, in *Design, Test, Integration & Packaging of MEMS/MOEMS* (2007)
20. M. Thompson, M. Li, D. Horsley, in *2011 IEEE 24th International Conference on Micro Electro Mechanical Systems (MEMS)* (2011), pp. 593–596. doi:10.1109/MEMSYS.2011.5734494
21. M. Li, V. Rouf, M. Thompson, D. Horsley, J. Microelectromech. Syst. **21**(4), 1002 (2012). doi:10.1109/JMEMS.2012.2196493

Chapter 3
Lorentz Force Magnetometers

The aim of this chapter is to review some main theoretical concepts related to MEMS that are used in the course of this book. After a brief introduction to dynamics of MEMS and to parallel-plates capacitive readout—a very popular technique for micromachined devices—and the related *pull-in* problem, this chapter deals with Lorentz force transduction principle applied to a micromachined structure. Magnetic field sensors are devices to measure the strength and/or direction of local magnetic field and they represent a fundamental block to integrate compass-based applications in mobile devices. Mechanical thermal noise, which sets the mechanical resolution of the device, is then introduced.

3.1 Capacitive Readout MEMS

3.1.1 Dynamics of MEMS

Capacitive readout MEMS are based on the measure of a capacitance variation due to the displacement of a suspended microscopic structure in presence of an external applied force. Regardless of transduction principle, a capacitive device consists of a mass which is free to move in one or more direction in the 3D space with respect to the substrate which it is anchored to through springs. Moving mass displacement is sensed capacitively: moving electrodes (also called *rotors* borrowing mechanical terminology) are mechanically anchored to the moving structure and fixed electrodes (as a consequence called *stators*) are anchored to the substrate. Figure 3.1 shows a differential capacitive sensing cell with a moving electrode anchored to a suspended shuttle (on the right) and forming a couple of capacitors with stators A and B.

C. Buffa, *MEMS Lorentz Force Magnetometers*,
DOI 10.1007/978-3-319-59412-5_3

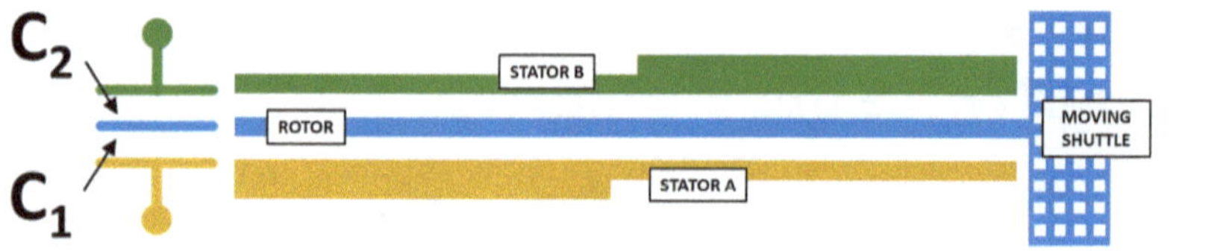

Fig. 3.1 Sketch of a typical differential capacitive sensing cell for a MEMS structure. Stators *A* and *B* are anchored to chip substrate and they form a differential capacitor with rotor

Fig. 3.2 Lumped parameters model of a 1-DOF spring-mass-damper system. Balance of forces acting on the microsystem

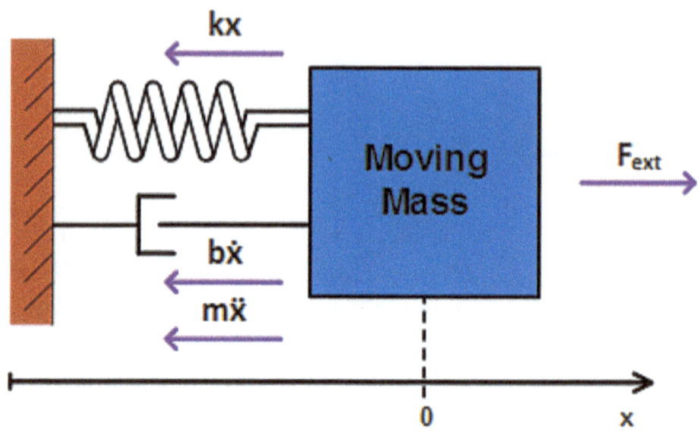

A microelectromechanical system can be modelled as a lumped parameters *spring-mass-damper* system [1], as shown in Fig. 3.2: a mass is connected via a *spring* to a fixed support, being pulled by an external force F_{ext}. It is also shown a *dashpot*, representing a mechanical damping element. All these three elements share the same displacement x with respect to a rest position. For sake of simplicity, a 1-axes model is considered now, neglecting secondary vibrating modes; this analysis can be easily extended to a 3-DOF system in an inertial frame of reference.

Applying Newton's second law of motion, stating that the net force on a body is equal to the product of acceleration and mass of the body:

$$\mathbf{F} = m \cdot \mathbf{a} \,, \tag{3.1}$$

the classical equation of motion describing the dynamics of a suspended micromachined structure can be derived.

The net applied force is due to the following components:

– An elastic force proportional to displacement x through *elastic stiffness*, k:

$$F_k = -k \cdot x \,; \tag{3.2}$$

– A viscous force which depends on the velocity through *damping coefficient*, b:

$$F_d = -b \cdot \dot{x} \,; \tag{3.3}$$

– An external applied force of different origins depending both on the transduction principle of the considered sensor (external acceleration, Lorentz force, Coriolis force, etc.) and on the readout technique (as shown in Sect. 3.1.2, electrostatic forces can occur).

Referring to Fig. 3.2, the balance of forces acting on the moving mass, along x axis can be mathematically written as:

$$m\ddot{x} + b\dot{x} + kx = F_{\text{ext}} , \tag{3.4}$$

where $\dot{x}$ and $\ddot{x}$ are, respectively, the first and second time derivative of the position x of moving mass m with respect to a rest position. Applying Laplace transform to Eq. (3.4):

$$ms^2 \cdot X(s) + bs \cdot X(s) + k \cdot X(s) = F_{\text{ext}}(s) , \tag{3.5}$$

displacement, $X(s)$, vs. external force, $F_{\text{ext}}(s)$, transfer function is obtained:

$$\frac{X(s)}{F_{\text{ext}}(s)} = \frac{1}{m} \cdot \frac{1}{s^2 + \frac{b}{m} \cdot s + \frac{k}{m}} = \frac{1}{m} \cdot \frac{1}{s^2 + \frac{\omega_r}{Q} \cdot s + \omega_r^2} . \tag{3.6}$$

Equation (3.6) highlights standard parameters of a second order system:

– Resonance frequency:

$$f_r = \frac{\omega_r}{2\pi} = \frac{1}{2\pi} \cdot \sqrt{\frac{k}{m}} ; \tag{3.7}$$

– Quality factor Q, a dimensionless parameter useful to characterize how over- or under-damped a MEMS resonator is. Equivalently, for large values, Q also characterizes resonator bandwidth Δf relative to its center frequency f_r and it is related to MEMS parameters according to:

$$Q = \frac{f_r}{\Delta f} = \frac{\omega_r \cdot m}{b} = \sqrt{\frac{k}{m}} \cdot \frac{m}{b} = \frac{\sqrt{k \cdot m}}{b} . \tag{3.8}$$

In order to evaluate system response to an external force, transfer function magnitude can be evaluated by replacing $s = j\omega$ in Eq. (3.6):

$$\begin{aligned} |T(j\omega)| &= \left| \frac{X(j\omega)}{F_{\text{ext}}(j\omega)} \right| \\ &= \frac{1}{m} \cdot \left| \frac{1}{\frac{k}{m} - \omega^2 + \frac{b}{m} \cdot j\omega} \right| \\ &= \frac{1}{m} \cdot \frac{1}{\sqrt{(\omega_r^2 - \omega^2)^2 + (\frac{\omega_r}{Q} \cdot \omega)^2}} . \end{aligned} \tag{3.9}$$

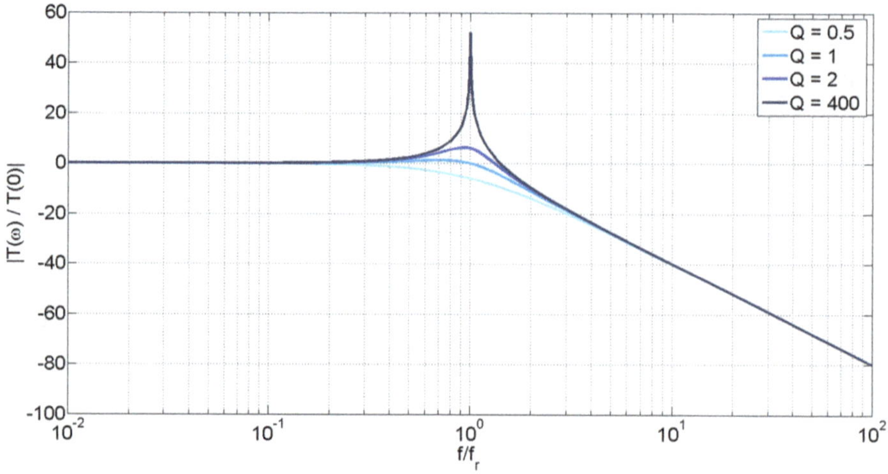

Fig. 3.3 Normalized magnitude of displacement to applied force transfer function vs. normalized frequency f/f_r

Figure 3.3 shows the normalized transfer function $|T(j\omega)|$ for four different values of Q (0.5, 1, 2, and 400). According to this chart, three different typical working conditions can be identified depending on frequency range:

- At frequencies lower than f_r—a typical working condition of accelerometers—displacement is proportional to external force through elastic stiffness with good agreement:

$$\omega \ll \omega_r \;\rightarrow\; \left|\frac{X(j\omega)}{F_{\text{ext}}(j\omega)}\right| \approx \frac{1}{k}\,; \tag{3.10}$$

- At frequencies higher than f_r, transfer function magnitude is attenuated by 40 dB per decade according to the theory of second order systems:

$$\omega \gg \omega_r \;\rightarrow\; \left|\frac{X(j\omega)}{F_{\text{ext}}(j\omega)}\right| \approx \frac{1}{m\cdot\omega^2}\,. \tag{3.11}$$

  This is not a working range of such an interest for MEMS applications since signal is strongly attenuated by the mechanical element itself.
- At resonance frequency, $f = f_r$, displacement is proportional to external force and it is amplified by a factor Q:

$$\omega = \omega_r \;\rightarrow\; \left|\frac{X(\omega)}{F_{\text{ext}}(\omega)}\right| \approx \frac{1}{bm} = \frac{Q}{m\omega_r^2} = \frac{Q}{k}\,. \tag{3.12}$$

By applying inverse Laplace transform, moving mass displacement with respect to external applied force can be obtained in the time domain:

$$x(t) = \frac{Q}{k} \cdot F_{\text{ext}}(t) \,. \tag{3.13}$$

Equation (3.13) shows that, at resonance frequency, moving mass displacement is directly proportional to external force and it has a "natural" magnification by a factor Q. Thus, working at f_r allows to increase device sensitivity, here intended as displacement variation with respect to external force variation. As detailed in the following, Lorentz force magnetometers usually operate in this condition. In Sect. 3.2 more details about Q and k are given considering the specific device architecture and package pressure.

3.1.2 *Electrostatic Forces: The Pull-in Phenomenon*

Due to their geometry and architecture, developed magnetometers can be modelled, from an electrical point of view, as a parallel-plates differential capacitor (more details are presented in Sect. 3.2): a "small" displacement (with respect to the distance between two electrodes) of the moving mass gives two ideally equal capacitance variations with opposite sign. Capacitive variations measurement is at the base of this readout technique.

At rest position, neglecting mechanical offset and fringing fields (a reasonable approximation when plates length and height are much longer compared to the distance between two electrodes), each of the two capacitances formed by a movable plate with the two stators is given by (Fig. 3.4a):

$$C_0 = \frac{\varepsilon_0 \cdot A_C}{g} \,, \tag{3.14}$$

where g is the distance between two electrodes (gap) at rest, A_C is overall sensing electrodes area, and ε_0 is vacuum dielectric constant.

In case of a displacement x, the two capacitances become (Fig. 3.4b):

$$\begin{aligned} C_1 &= \varepsilon_0 \cdot \frac{A_C}{(g+x)} \\ C_2 &= \varepsilon_0 \cdot \frac{A_C}{(g-x)} \,. \end{aligned} \tag{3.15}$$

Because of opposite charges on two plates forming a capacitor there is a force of attraction between plates which is commonly neglected in fixed-plates electrical capacitors. This charge is always present whenever the capacitor is charged and, in case of at least one movable plate, the inclusion of this mechanical force becomes essential.

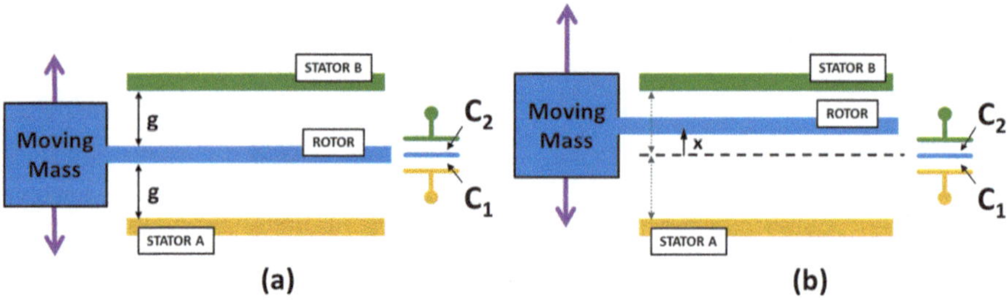

Fig. 3.4 MEMS modelled as a variable differential parallel-plates capacitor. (**a**) At rest position; (**b**) at a displacement x with respect to rest position

Fig. 3.5 A single-ended capacitor with a moving electrode (*in blue*) and a stator (*in yellow*) is biased at a constant voltage V

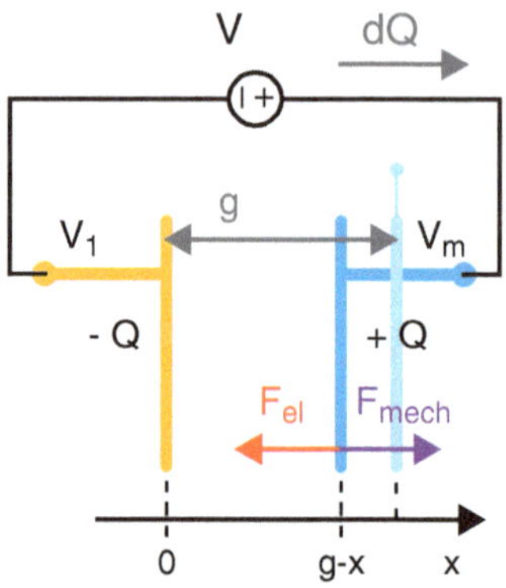

A single capacitor is now considered, biased with a voltage generator which keeps the potential between the two electrodes at a constant voltage V. This capacitor is formed by a moving electrode, V_m, and a stator, V_1, which is anchored to a substrate, as shown in Fig. 3.5. After applying a constant voltage, the moving electrode moves in the negative x direction because of the attraction of electrostatic forces; the mechanical spring opposes the movement of the rotor and so the mechanical force F_{mech} performs a negative work on the system. Moreover, if the distance between electrodes varies, capacitance value also changes and, on a fixed biasing voltage, so does the charge on electrodes. In this example, the moving electrode gets closer to the anchored electrode resulting in a capacitance increase. Moreover the voltage generator performs a positive work to transfer positive charges onto electrode V_m and the net energy stored in the capacitor increases. Because of conservation of energy:

$$dE_C = dW_{\text{el}} - dW_{\text{mech}} \, , \tag{3.16}$$

where W_{mech} is mechanical work due to rotor displacement, W_{el} is the work associated to electrostatic force, and E_C is energy stored in the capacitor. The latter can be written as:

$$E_C = \frac{1}{2}C \cdot V^2 , \tag{3.17}$$

where C is the capacitance and V voltage across the capacitor. Differentiating Eq. (3.17) and considering that there are not any potential variations between electrodes (a DC voltage generator is connected to this capacitor), it is obtained:

$$dE_C = \frac{1}{2}V^2 \cdot dC . \tag{3.18}$$

In order to calculate electrical work:

$$dW_{\mathrm{el}} = V \cdot dQ , \tag{3.19}$$

the charge on a capacitor $Q = V \cdot C$ is differentiated (always under the assumption of a constant voltage):

$$dQ = V \cdot dC . \tag{3.20}$$

This relationship states that the charge stored in a capacitor changes according to the value of the capacitor so that the potential across the capacitor does not change.

By replacing Eq. (3.20) in Eq. (3.19), the electrical work is obtained:

$$dW_{\mathrm{el}} = V^2 \cdot dC . \tag{3.21}$$

Finally, mechanical work component is the product of displacement and the force which generates that displacement:

$$dW_{\mathrm{mech}} = F_{\mathrm{mech}} \cdot dx . \tag{3.22}$$

By replacing Eqs. (3.22), (3.21), and (3.18) in Eq. (3.16), the final relationship is derived:

$$\frac{1}{2}V^2 \cdot dC = V^2 \cdot dC - F_{\mathrm{mech}} \cdot dx , \tag{3.23}$$

resulting in:

$$F_{\mathrm{mech}} \cdot dx = \frac{1}{2} \cdot V^2 \cdot dC . \tag{3.24}$$

By differentiating the capacitance with respect to the displacement:

$$dC = \frac{\varepsilon_0 \cdot A_C}{(g - x)^2} \cdot dx , \tag{3.25}$$

and combining (3.25) with (3.24) the resulting mechanical force, which is at equilibrium equal to the electrostatic force, is:

$$F_{\text{el}} = F_{\text{mech}} = \frac{1}{2}\varepsilon_0 \cdot A_C \cdot \frac{V^2}{(g-x)^2}\ . \tag{3.26}$$

Finally, MEMS motion equation must be refined including an electrostatic component:

$$m\ddot{x} + b\dot{x} + kx + F_{\text{el}} = F_{\text{ext}}\ . \tag{3.27}$$

Depending on operating conditions and values of parameters, balance of forces can be dominated by a specific component and Eq. (3.27) can have different solutions: either stable or unstable. In case of voltage controlled parallel-plates capacitors, an important behavior, called *pull-in*, might happen: at some critical voltage the system goes unstable and the gap collapses to zero. Following a stability analysis of the equilibrium between elastic force and electrostatic force, it can be shown that pull-in occurs at [1]:

$$x_{\text{pi}} = \frac{g}{3} \tag{3.28}$$

and, with this value of displacement, the equilibrium voltage is:

$$V_{\text{pi}} = \sqrt{\frac{8}{27} \cdot \frac{k \cdot g^3}{\varepsilon_0 \cdot A_C}}\ . \tag{3.29}$$

Equation (3.29) provides pull-in voltage for a single-ended parallel-plates capacitor.

This treatment can be extended to the case of a differential capacitor where V_1 is stator A potential, V_2 is stator B potential, and V_m is moving mass potential. Resulting electrostatic forces act in opposite directions on the moving electrode and they are (Fig. 3.6)

$$F_{\text{el},1} = -\frac{1}{2}\varepsilon_0 \cdot A_C \cdot \frac{(V_1 - V_m)^2}{(g+x)^2}\ , \tag{3.30}$$

$$F_{\text{el},2} = \frac{1}{2}\varepsilon_0 \cdot A_C \cdot \frac{(V_m - V_2)^2}{(g-x)^2}\ . \tag{3.31}$$

MEMS motion equation including electrostatic components of a differential capacitor results in:

$$m\ddot{x} + b\dot{x} + kx + F_{\text{el},1} - F_{\text{el},2} = F_{\text{ext}}\ . \tag{3.32}$$

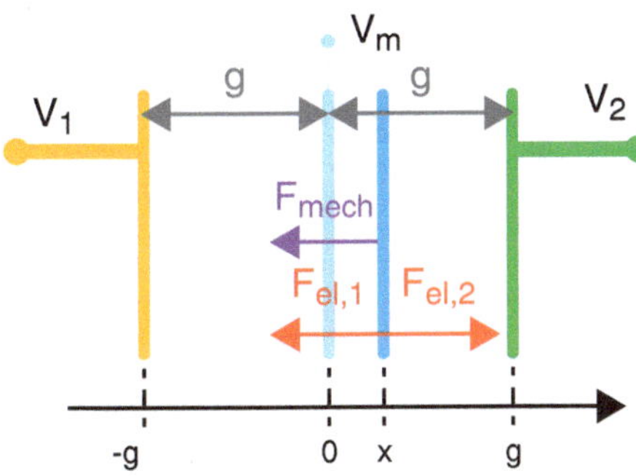

Fig. 3.6 A differential capacitor with a moving electrode (*in blue*) and stators (*in green and yellow*)

For a differential capacitor, pull-in voltage occurs at:

$$V_{\text{pi,diff}} = \sqrt{\frac{k \cdot g^3}{\varepsilon_0 \cdot A_C}}. \tag{3.33}$$

3.2 *Z*-axis Sensitive Magnetometer

Figure 3.7 illustrates a simplified representation of *Z*-axis parallel-plates MEMS magnetometers. Two beams with a length L constitute the suspending element of a central shuttle, which forms with a set of fixed stators two differential parallel-plates sensing capacitors C_1 and C_2. For an alternated current $I(t) = i_r \cdot \sin(2\pi \cdot f_r \cdot t)$ flowing through the springs as depicted at the device resonance frequency f_r, the structure is subject to the Lorentz force $F_L = I(t)/2 \cdot L \cdot B(t)$ on each spring and thus to a displacement amplified through the quality factor Q:

$$x(t) = \frac{Q}{2k} \cdot 2 \cdot F_L(t) = \frac{Q \cdot I(t) \cdot L \cdot B(t)}{2 \cdot k}, \tag{3.34}$$

k being the total device stiffness, i.e., four times the elastic stiffness of each suspending beam [2]:

$$k = 4 \cdot k_{\text{beam}} = 4 \cdot \frac{12 \cdot E_Y \cdot I_{\text{Inertia}}}{L_s^3}, \tag{3.35}$$

where L_s is the length of the beam from the substrate anchorage to the central shuttle, E_Y is the modulus of elasticity of the beam material (polysilicon), and I_{Inertia} is the area moment of inertia about the centroidal axis of the beam cross-section.[1] It is [2]:

[1]This unusual symbol is used to avoid misleading this quantity with current, denoted in all the manuscript with the symbol I.

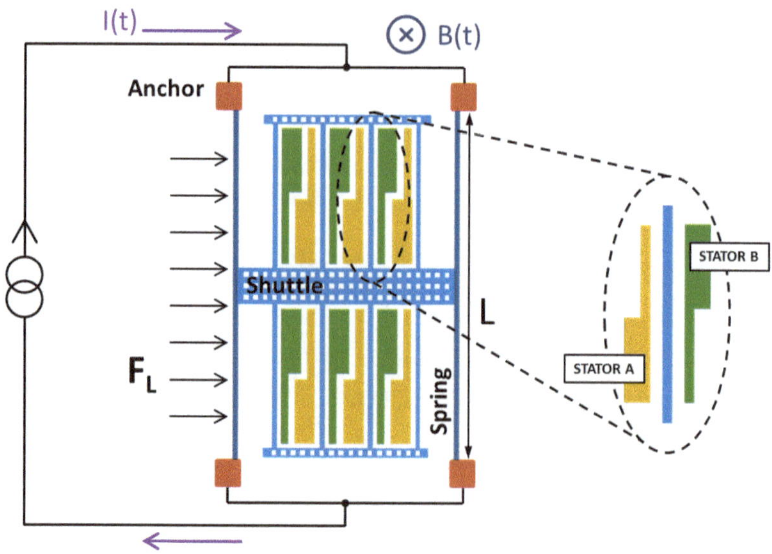

Fig. 3.7 Schematic view of a *Z*-axis parallel-plates MEMS magnetometer. The suspended mass is subject to a Lorentz force in presence of a magnetic field; the corresponding displacement can be sensed through differential capacitors

$$I_{\text{Inertia}} = \frac{1}{12} \cdot H \cdot W^3 \,, \tag{3.36}$$

where H is process height and W springs width.

Combining Eqs. (3.35) and (3.36) the device total elastic stiffness turns out to be:

$$k = 4 \cdot \frac{E_Y \cdot H \cdot W^3}{L_s^3} \,. \tag{3.37}$$

The factor 2 at the denominator of Eq. (3.34) accounts for the fact that the Lorentz force is distributed across the whole length L and not concentrated on the shuttle.

Referring to a differential parallel-plates cell like the one shown in Figs. 3.4 and 3.7, capacitance variation can be written as follows:

$$\begin{aligned} \Delta C(t) &= C_2(t) - C_1(t) \\ &= \varepsilon_0 \cdot \frac{A_C}{(g - x(t))} - \varepsilon_0 \cdot \frac{A_C}{(g + x(t))} \\ &= 2\varepsilon_0 \cdot \frac{A_C \cdot x(t)}{(g^2 - x^2(t))} \end{aligned} \tag{3.38}$$

In approximation of small displacements with respect to rest position, which means $x << g$, Eq. (3.38) can be approximated to:

$$\begin{aligned} \Delta C &\approx 2C_0 \cdot \frac{x(t)}{g} \\ &= 2\varepsilon_0 \cdot A_C \cdot \frac{x(t)}{g^2} = 2\varepsilon_0 \cdot L_C \cdot N \cdot H \cdot \frac{x(t)}{g^2} \end{aligned} \tag{3.39}$$

where ε_0 is the electrical permittivity inside the package (assumed as that of vacuum) and A_C is the overall facing area of the parallel plates, that is equal to the product of the process height H, the length of each parallel-plate L_C, and the number of differential sensing cells N.

From Eq. (3.39) combined with Eq. (3.34) it can be seen that this kind of sensors work in amplitude modulation mode: an excitation is provided at resonance frequency, f_r, and the magnitude of an external magnetic field modulates the amplitude of capacitance variation. The detection of this capacitance variation contains the information about magnetic field variation. The described operating principle is valid as long as the frequency of external magnetic field is lower than device mechanical bandwidth. On the other hand, if magnetic field frequency is out of device mechanical bandwidth no signal can be detected.

3.3 Mechanical Sensitivity at Fixed Resonance Frequency

In this section and in the following of the manuscript, the term *mechanical sensitivity* refers to the sensitivity of the mechanical element which transduces the input quantity, an external magnetic field, into an intermediate output, a capacitance variation ($\Delta B \rightarrow \Delta C$). Then, readout electronics is coupled to convert this capacitance variation into a system output voltage ($\Delta C \rightarrow \Delta V$). The overall system sensitivity is given in terms of output voltage per input magnetic field ($\Delta B \rightarrow \Delta V$).

Among all specifications of Table 2.1, the requirement of a resonance frequency which is out of the audio bandwidth, and so the requirement to fix a resonance frequency in the design of new devices, leads to a completely new design approach. In the literature no specific design approaches are shown concerning the resonance frequency of the device, even though some of proposed devices work out of the audio bandwidth. In the present work, theoretical analyses and modelling are based on the idea that resonance frequency is fixed and set right beyond the audio bandwidth. A resonance frequency above 20 kHz is enough to avoid acoustic coupling to other sensors embedded in mobile devices like microphones, MP3 players, speakers, etc.

The mechanical sensitivity $\Delta C/\Delta B$ of this magnetometer architecture, defined as the differential capacitance variation per magnetic field variation, can be written as a function of the resonance frequency as follows:

$$
\begin{aligned}
\frac{\Delta C}{\Delta B} &= 2\varepsilon_0 \cdot L_C \cdot N \cdot H \cdot \frac{Q \cdot I(t) \cdot L}{g^2 \cdot 2 \cdot k} \\
&= \varepsilon_0 \cdot L_C \cdot N \cdot H \cdot \frac{\omega_r \cdot m \cdot I(t) \cdot L}{g^2 \cdot k \cdot b} \\
&= \varepsilon_0 \cdot L_C \cdot N \cdot H \cdot \frac{\omega_r \cdot m \cdot I(t) \cdot L}{g^2 \cdot \omega_r^2 \cdot m \cdot b} \\
&= \varepsilon_0 \cdot L_C \cdot N \cdot H \cdot \frac{I(t) \cdot L}{g^2 \cdot 2\pi \cdot f_r \cdot b} .
\end{aligned}
\tag{3.40}
$$

In the formula above, b is the damping coefficient. It is known that this coefficient depends linearly on the pressure in the so-called transition regime, where the free-molecule flow applies [3]. This means that the damping is mostly caused by the collisions between the structure and the gas molecules, with a negligible interaction between the molecules themselves. As a consequence, as the largest number of collisions occurs between the parallel plates and the squeezed fluid, the coefficient b is proportional both to the area A_C and to the number N, to a first approximation, through a normalized damping coefficient per unit area b_{area}:

$$
b = 2 \cdot b_{\text{area}} \cdot N \cdot L_C \cdot H \tag{3.41}
$$

(the factor 2 accounts for the differential configuration). By replacing Eq. (3.41) in Eq. (3.40), it thus turns out that the sensitivity defined above is independent on the number of parallel-plate cells:

$$
\frac{\Delta C}{\Delta B} = \frac{\varepsilon_0 \cdot I(t) \cdot L}{4\pi \cdot g^2 \cdot f_r \cdot b_{\text{area}}} . \tag{3.42}
$$

Looking at the derived formula, some considerations can be made about possible optimization of the design:

1. a lower value of gap g increases sensitivity but the minimum air gap is set by the technology. Choosing the minimum value means increasing the variation of rest capacitor value and so worsening the matching among cells. At system level this increases production spread from part to part, making calibration necessary [4, 5];
2. the resonance frequency f_r is set in the range 20–30 kHz to avoid acoustic interference, as commonly done for gyroscopes [6, 7];
3. the damping coefficient per unit area b_{area} is constrained by the packaging pressure and by the open-loop bandwidth of the device, $\text{BW} = f_r/(2Q)$ (see Fig. 3.8). As an example, in order to have a bandwidth of 50 Hz the quality factor should not exceed a value around 200–300;
4. the spring length L and the driving current $I(t)$ can be increased to increase the sensitivity. In the first case, in order not to change the elastic stiffness of the device and so to cope with point 2 above (supposing that the increase of the mass

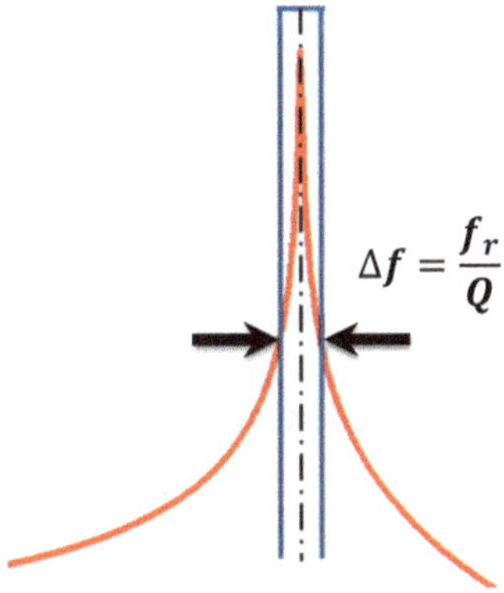

Fig. 3.8 Close-up of a typical MEMS transfer function (*red curve*) at resonance frequency, highlighting the full width at half maximum (FWHM) equal to $\Delta f = f_r/Q$. The mechanical bandwidth of the device, defined as BW $= \Delta f/2 = f_r/2Q$, sets the maximum detectable bandwidth of the input signal which can be filtered with a low-pass filter (*in blue*, an ideal one) after demodulation

is negligible) springs width must be adjusted according to $dW/W = dL/L$. This means that their resistance does not change and an increase of springs length does not cost a higher power dissipation. Technology constraints are the only limit.

On the other hand, the driving current can be increased at the cost of an increased power dissipation by Joule effect in the springs, $P_{\text{Joule}} \propto I^2(t)$.

Most important, the found independence of the sensitivity on the number N of sensing cells is relevant as it suggests the design of a Z-axis MEMS magnetometer which can be considered one-dimensional. Being smaller in terms of area, it can fit in a lateral side of the same package of the three-axes gyroscope, which shares with the magnetometer the same pressure requirement (it can also be combined in the same package with other devices depending on which sensors are integrated in the same IMU). While several examples of parallel-plate magnetometers in the literature show a large number of parallel plates [8, 9], a recent device proposed in [10] is based on a small number of sensing cells even if it uses comb fingers and no discussion is presented about the impact of the combs number.

External accelerations or vibrations, typically at frequencies lower than a few kHz [11], can be filtered and do not change significantly the operating point of the magnetometer, thanks to the high resonance frequency (acceleration induced displacements are indeed proportional to the inverse of f_r^2: $x = a/\omega_r^2$).

3.3.1 Damping Modelling

The technology used to fabricate devices, which are shown in this manuscript, is Thick Epitaxial Layer for Micro-gyroscopes and Accelerometers (ThELMA).

This is a surface micromachining process by STMicroelectronics, which combines (variably) thick and thin polysilicon layers for structures and interconnections. It is based on the deposition of a sacrificial oxide layer, on which a 15 μm thick epitaxial layer of polysilicon is first grown, then micromachined through deep reactive ion etching, and finally released through an HF etching. More details on the process can be found in Sect. 5.2.1. The packaging is made through a wafer–wafer bonding at a nominal pressure around 1 mbar. For those devices which require a large quality factor (gyroscopes, magnetometers) a getter material is used within the sealed chamber, to keep the low pressure constant during the device lifetime.

At the scales of interest in microsystems, rarefaction effects in the gas flow always play a key role. These are conventionally quantified by means of the Knudsen number, defined as $Kn = \lambda/\ell$, where ℓ is a characteristic length of the flow and λ is the mean free path of gas molecules. Based on the Knudsen number, the gas flow can be qualitatively classified as in continuum ($Kn < 0.01$), slip ($0.01 < Kn < 0.1$), transition ($0.1 < Kn < 10$), or free-molecule ($Kn > 10$) regime. If we take for ℓ the typical gap between parallel plates (e.g., $\ell = 2$ μm) and use $\lambda = 69$ μm (at a pressure $p = 1$ mbar), then we find that the designed magnetometers fall within the free-molecule flow regime ($Kn \approx 35$).

In the above-mentioned working conditions, gas dissipation was computed both employing custom statistical tools, like the *Test Particle Monte Carlo Method*, and deterministic procedures for the solution of the collision-less Boltzmann model based on integral equations (see [12, 13]). Since the largest contribution to dissipation comes from local squeeze effects in between the parallel plates, the analysis can be conveniently performed by assuming perfect decoupling between the different plate/stator units. The damping force for a single unit is computed using the algorithm and then multiplied by the number N of shuttle plates. The resulting damping coefficient per unit area turns out to be from the model $b_{\text{area}} = 5.8\ \text{kg}/(\text{s}\,\text{m}^2)$ at the nominal pressure of 1 mbar.

To discuss possible side effects for this model, one should consider the parallel-plates dimensions. The height, $H = 15$ μm, is fixed by the technology. The gap is chosen to be the minimum ($g \approx 2$ μm) so that the sensitivity is maximized: indeed even if the damping coefficient per unit area b_{area} is itself an increasing function with respect to $1/g$, on the whole the sensitivity (which is proportional to the factor $1/(g^2 \cdot b_{\text{area}})$) increases with decreasing g. Finally, the sensitivity does not depend on the length of the parallel-plates L_C (see Eq. (3.42)). Yet the open-loop bandwidth ($f_r/2Q$) is a function of the damping coefficient, and in turn of the length L_C. Therefore it was chosen to have an overall parallel-plate length of 332 μm, which from simulations guarantees a maximum Q around 450 at the reference pressure (and thus a bandwidth at least >30 Hz). More details about simulations results and implemented devices geometry are given in Sect. 4.2. Therefore, for sake of simplicity, as the length and the height of each parallel-plate cell are far larger than the gap dimension, the algorithm for the calculation of the damping coefficient neglects side effects.

3.4 Thermo-Mechanical Noise

Micromechanical devices are affected by thermal noise like all dissipative systems. In particular, dimensions scaling is attractive for a higher density integration, but small moving parts become more susceptible to mechanical noise due to molecular movement. Especially in sensors targeted for very low signals applications, mechanical noise may be a limiting factor.

In gas damped systems, like MEMS working either at ambient pressure or in a package at a lower pressure, mechanical noise is mainly due to random paths of molecules which hit the suspended structure. The result of this statistic process is an unwanted random displacement of the moving mass which is nevertheless detected by position sense interface.

3.4.1 Intrinsic Resolution

The minimum detectable magnetic field is determined by all the noise sources, i.e., both the intrinsic thermo-mechanical noise of the device and the readout electronics noise. The first and unavoidable contribution is due to thermal movement of gas particles inside the package hitting the moving mass and thus resulting in a random vibration. Thermo-mechanical noise expressed in terms of power spectral density of the noise force acting on a device is [14, 15]:

$$S_{F_n} = 4k_{\mathrm{B}}Tb\,, \tag{3.43}$$

T being the absolute temperature and k_{B} the Boltzmann constant. It should not surprise that the resulting expression for mechanical noise is very similar to Johnson noise in resistors, $S_{V_n} = 4k_{\mathrm{B}}TR$, as they both have the same physical origin, dissipation. Equation (3.43) can be converted into a displacement noise density according to the transfer function of the MEMS (we assume here Q/k at resonance, indeed the Brownian noise acts mostly on the suspended shuttle and not as a distributed force on the springs). A consistent assumption is to consider a filter in the readout chain that sets the maximum readout bandwidth at $\mathrm{BW} = \Delta f/2 = f_r/(2Q)$, which corresponds to the -3 dB mechanical bandwidth of the device, as shown in Fig. 3.8. The intrinsic signal to noise ratio (SNR) then results:

$$\mathrm{SNR_{int}} = \frac{\frac{Q \cdot I(t) \cdot L \cdot B}{2k}}{\sqrt{S_{F_n} \cdot \frac{Q^2}{k^2} \cdot \frac{f_r}{2Q}}}\,. \tag{3.44}$$

Conventionally accepting an SNR $= 1$, the minimum magnetic field which can be detected (without considering electronics noise) turns out to be:

$$B_{\text{min}_{\text{int}}} = \frac{8 \cdot \text{BW}}{I \cdot L} \cdot \sqrt{k_B \cdot \pi \cdot T \cdot m}\,, \tag{3.45}$$

where m is the device mass and BW is here set equal to $f_r/2Q$.

The intrinsic resolution finally results to be independent from the number of sensing cells N, like the sensitivity. That being stated, the choice of an optimum number of sensing cells must take into account that the minimum detectable field decreases with the mass square root and so a lighter device is preferable. On the other hand, long springs can be desirable to improve both the sensitivity and the resolution but the intrinsic power dissipation and the maximum suspended structure length set a trade-off.

It is interesting to apply the proposed theory to some scientific works shown in Sect. 2.4. Assuming the effective length L for the Lorentz force calculation roughly equal to the maximum device dimensions, for [16] it can be evaluated an intrinsic magnetic field noise density of 180 $\text{nT}/\sqrt{\text{Hz}}$, quite in line measured value by the authors (140 $\text{nT}/\sqrt{\text{Hz}}$). Similarly, for the device described in [8], the here proposed model predicts an intrinsic magnetic field noise density of 113 $\text{nT}/\sqrt{\text{Hz}}$, slightly better than the result measured by the authors (200 $\text{nT}/\sqrt{\text{Hz}}$), which, however, states that there are electronic noise limitations in their measurements. For [17] the impossibility to estimate the device mass, and above all the presence of smart multiple current paths on each suspended spring (allowed by the process there used to deposit and isolate metal paths from the structural layer) make it difficult a direct application of Eq. (3.45).

References

1. S.D. Senturia, *Microsystem Design* (Kluwer Academic, Norwell, 2001)
2. W.C. Young, R.G. Budynas, *Roark's Formulas for Stress and Strain*, 7th edn. (McGraw-Hill, New York, 2001)
3. L. Mol, L. Rocha, E. Cretu, R. Wolffenbuttel, J. Micromech. Microeng. **19**(7), 074021 (2009). http://stacks.iop.org/0960-1317/19/i=7/a=074021
4. F. Camps, S. Harasse, A. Monin, in *IEEE International Conference on Electro/Information Technology, 2009. eit 09* (2009), pp. 217–221. doi:10.1109/EIT.2009.5189614
5. G.A. Aydemir, A. Saranli, Measurement **45**(5), 1210 (2012). doi:10.1016/j.measurement.2012.01.015. http://www.sciencedirect.com/science/article/pii/S0263224112000309
6. R. Dean, S. Castro, G. Flowers, G. Roth, A. Ahmed, A. Hodel, B. Grantham, D. Bittle, J. Brunsch, IEEE Trans. Ind. Electron. **58**(7), 2591 (2011). doi:10.1109/TIE.2010.2070772
7. R. Dean, G. Flowers, A. Hodel, G. Roth, S. Castro, R. Zhou, A. Moreira, A. Ahmed, R. Rifki, B. Grantham, D. Bittle, J. Brunsch, in *IEEE International Symposium on Industrial Electronics, 2007. ISIE 2007* (2007), pp. 1435–1440. doi:10.1109/ISIE.2007.4374812
8. H. Emmerich, M. Schöfthaler, IEEE Trans. Electron Devices **47**(5), 972 (2000). doi:10.1109/16.841228
9. M. Thompson, D. Horsley, in *2009 IEEE Sensors* (2009), pp. 992–995. doi:10.1109/ICSENS.2009.5398216
10. M. Thompson, M. Li, D. Horsley, in *2011 IEEE 24th International Conference on Micro Electro Mechanical Systems (MEMS)* (2011), pp. 593–596. doi:10.1109/MEMSYS.2011.5734494

11. R. Dean, A. Anderson, S. Reeves, G. Flowers, A. Hodel, IEEE Trans. Ind. Electron. **58**(7), 2697 (2011). doi:10.1109/TIE.2010.2076310
12. A. Frangi, A. Ghisi, L. Coronato, Sens. Actuators, A **149**(1), 21 (2009). doi:10.1016/j.sna.2008.09.018. http://www.sciencedirect.com/science/article/pii/S0924424708004810
13. R. Leung, T. Thurber, W. Ye, Microfluid. Nanofluid. **11**(6), 753 (2011)
14. T. Gabrielson, IEEE Trans. Electron Devices **40**(5), 903 (1993). doi:10.1109/16.210197
15. J. Vig, Y. Kim, IEEE Trans. Ultrason. Ferroelectr. Freq. Control **46**(6), 1558 (1999). doi:10.1109/58.808881
16. M. Li, V. Rouf, M. Thompson, D. Horsley, J. Microelectromech. Syst. **21**(4), 1002 (2012). doi:10.1109/JMEMS.2012.2196493
17. J. Kyynäräinen, J. Saarilahti, H. Kattelus, A. Kšrkkäinen, T. Meinander, A. Oja, P. Pekko, H. Seppä, M. Suhonen, H. Kuisma, S. Ruotsalainen, M. Tilli, Sens. Actuators, A **142**(2), 561 (2008). doi:10.1016/j.sna.2007.08.025

Chapter 4
System Design

4.1 System Behavioral Model

This section presents a behavioral model for MEMS magnetometers at a system level, which is implemented using Simulink® by Mathworks®, a block diagram environment for multidomain dynamic simulations. One of the main reasons for MEMS rapid growth is the possibility to combine actuating and sensing mechanical elements together with signal conditioning electronics in a system which is getting more and more complex. Simulations are fundamental to ensure the proper functioning and to comply with design constraints. Moreover, from an industrial point of view, a design flow aided by system models helps to reduce time-to-market and prototyping costs. Magnetic field sensing system modelling allows to investigate the working principle and predict resonator's behavior, to evaluate effects of process spreads and changes of mechanical structure, to study driving and readout mechanisms, and to evaluate system performance before implementation [1].

Magnetometer system modelling has been performed following a bottom-up approach in order to divide the model complexity by adding separate subsystems. This approach leads to develop in details the base elements of the system and then they are connected together to form larger subsystems and then to assemble a top-level cell. This model incorporates both mechanics and electronics. The resulting model features a highly detailed implementation of the system, providing a time discretization which is fine enough to appreciate signals during system start-up and transients. Despite such a complexity, simulation runtime is in order of tens of minutes for about hundreds ms simulation,[1] making system level modelling even more attractive for product development.

[1]Simulation runtime is estimated on an up-to-date performing hardware.

C. Buffa, *MEMS Lorentz Force Magnetometers*,
DOI 10.1007/978-3-319-59412-5_4

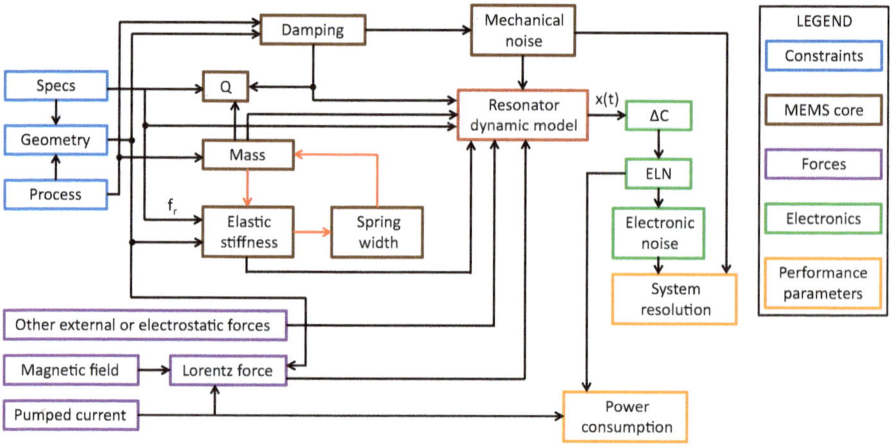

Fig. 4.1 Simplified block diagram of the complete magnetometer Simulink® model

A simplified block diagram of magnetometer system is shown in Fig. 4.1; the schematic includes a resonator block, which is the core of the whole system, in addition to blocks for specifications, geometrical and process parameters description (input data), Lorentz and other electrostatic or external forces, elastic stiffness, damping, and mass. Main output of the block which solves motion equation is moving mass displacement which is fed to a block for capacitance variation evaluation. On the base of this variable capacitance variation the resulting output voltage is calculated. The model is developed according to the new approach proposed in theory (Sect. 3.2): the resonance frequency of the simulated device is fixed according to specifications and it is an input of the model. Mass is calculated in a sub-block of the system starting from a detailed geometrical description of the device; mass depends on number of sensing cells, N, an input parameter, and also on springs mass. But the exact mass of springs depends on the width of springs themselves which is calculated using resonance frequency and total effective mass. In order to solve this loop (see red arrows in Fig. 4.1) the software starts with an initial guess about the width of springs which is used for an initial estimation of the effective mass. Then the elastic stiffness is calculated using input value of resonance frequency (fixed) and the obtained estimation of mass. Inverting Eq. (3.37) a new springs width can be calculated:

$$W = \sqrt[3]{\frac{k}{4} \cdot \frac{L_S^3}{E_Y \cdot H}}, \tag{4.1}$$

where E_Y is the modulus of elasticity and L_s is the length of each suspending beam, as already mentioned in Sect. 3.2.

The process is iterated until springs dimensions satisfy the loop condition: the refinement of spring width W from the previous iteration should be lower than a preset error, which is typically chosen in the order of technology tolerances. Then the software solves resonator equation, calculate capacitance variation and related electronics noise and gives output.

4.1.1 Geometrical Data, Specifications, and Process Parameters

Among system inputs, there are blocks for geometrical description of the device, process parameters according to the used technology and specifications. As far as geometry section is concerned, the most important dimensions which can be controlled are the drawn minimum allowed gap, lengths of springs whose maximum value is set by technology to avoid buckling (collapse of a suspended structure onto the substrate), length, width, and number of capacitive electrodes which have an impact on device dimensions. As far as process parameters are concerned, there are the height of the structural layer, typical process overetch, and physical parameters like polysilicon density and Young modulus; these technological parameters are estimated by Corigliano et al. [2]. The developed software turns out to be extremely versatile because simulations considering different technology parameters or a redesign of the geometry can be performed only by changing the value in this specific subsystem.

4.1.2 Damping Modelling

Damping effects are integrated in the system model on the base of study and theory proposed in Sect. 3.3. Equation (3.41) is solved by Simulink® with a value of $b_{\text{area}} = 5.8\ \text{kg}/(\text{s}\,\text{m}^2)$ estimated at a pressure of 1 mbar for a drawn gap of 1.4 μm and an overetch of 0.7 μm. The term *overetch* refers to the extra amount of silicon with respect to design dimensions which is removed during etching process step.

4.1.3 Mathematical Modelling of the Resonator

The core of the model is a block to solve MEMS motion equation:

$$m\ddot{x} + b\dot{x} + kx + F_{\text{el},1} - F_{\text{el},2} = F_{\text{ext}}\ . \tag{4.2}$$

All parameters involved in Eq. (4.2) are determined by analytical methods in embedded subsystems, already introduced and some of them are described more in details in the following.

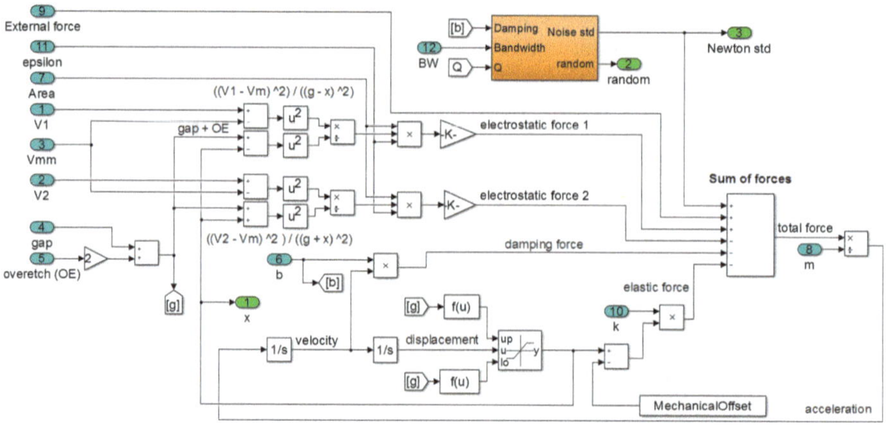

Fig. 4.2 Simulink® block diagram of sub-block solving MEMS motion equation

The equation of motion which represents the parallel-plates resonator is implemented in Simulink® as a transfer function, as shown in Fig. 4.2. The resolution[2] of this equation corresponds to a sum of forces. Dividing this total force at any instant by the mass gives the acceleration of the shuttle. The integration of the acceleration, represented in the $1/s$ block (in the Laplace domain an integration corresponds to divide by s) gives the speed and after a successive integration, the displacement of the shuttle can be found at any particular instance. A saturation dynamic block is used as a mathematical implementation of a physical stopper and it helps to avoid solving equations when the distance between the moving and the fixed electrodes tends to zero and the electrostatic force diverges. Damping, mass, and elastic stiffness, here simply used as multiplication coefficients, are calculated in separate subsystems. In Fig. 4.2 blocks for the evaluation of the electrostatic forces are shown too.

Brownian thermo-mechanical noise, separately calculated with estimated damping coefficient, is added to sum of forces. The noise block calculates the root mean square of the Brownian noise over the bandwidth of the system starting from the damping value and, through the use of a random number generator (provided in Simulink® libraries), it is converted in a random force with a Gaussian distribution.

The main output of the mechanical resonator subsystem is the displacement of the shuttle which is fed to capacitance calculation block and it represents one of the system global outputs for further processing and monitoring using MATLAB® command line.

[2]The word *resolution* is here used with its mathematical meaning and it must not be mistaken with its common meaning in sensors and instrumentation field, which is the smallest change in the physical quantity which gives a response in the measurement.

4.1.4 Lorentz Force

The Lorentz force is calculated in a subsystem the inputs of which are springs dimensions, pumped current, and external magnetic field. Springs dimensions are calculated in another subsection of the system (satisfying resonance frequency specification) whereas the current and the magnetic field can be set in the execution file; both static and time dependent magnetic field waveform can be used. This block output represents one of the elements of external forces array in the solution of motion equation.

4.1.5 System Model Outputs

After model resolution on a time interval which can be chosen by the user according to the required study, a lot of variables are available as system output signals; they are evaluated at any instant of the simulation time and they can be used in the MATLAB® workspace for further processing. Here is the list of the most significant variables:

- displacement of moving shuttle;
- capacitance value at rest, single-ended and differential capacitance variations;
- quality factor and damping coefficient;
- mass (both springs and shuttle) and resonance frequency;
- elastic stiffness and springs dimensions.

After several sets of simulations for system optimization, the model is used to simulate magnetometers with a fixed resonance frequency (set by the user) and different numbers of sensing cells. Devices are implemented on the base of results provided by the developed model; in particular springs dimensions and number of sensing cells vary for each structure.

Section 4.2 reports main results of system level simulation and in Chap. 5 are reported fabrication process and layouts.

4.2 Simulations Results and Devices Design

Results of simulations are presented in this section. Main used parameters are: beam length $L = 870$ μm (this value was chosen, at the time of first design, according to the maximum suspended beam length allowed by technology design rules in order to avoid buckling), length of parallel-plate electrode $L_C = 332$ μm (for this electrode length value fringe field effects can be neglected with good approximation; moreover this value affects damping coefficient as detailed in Sect. 3.3.1), process height $H = 15$ μm, thickness of parallel-plate electrode $W_C = 3$ μm (a minimum

Table 4.1 Simulation results for parallel-plates magnetometers with different numbers N of sensing cells and for a pumped current of 200 μA_{rms} at resonance

Data	$N=2$	$N=4$	$N=6$	$N=8$	$N=10$	$N=16$	$N=32$
k (N /m)	14.3	17.6	20.9	24.1	27.3	36.8	60.5
Q	839	518	409	354	321	270	222
C_0 (fF)	42	84	126	168	210	335	670
W_{spring} (μm)	5.02	5.39	5.70	5.98	6.23	6.88	8.12
m_{spring} (nkg)	0.297	0.319	0.338	0.354	0.369	0.408	0.481
m_{tot} (nkg)	0.452	0.558	0.661	0.763	0.863	1.163	1.912
Area (m^2)	10e−9	20e−9	30e−9	40e−9	50e−9	80e−9	160e−9
b (Ns/m)	9.57e−8	1.91e−7	2.87e−7	3.82e−7	4.79e−7	7.66e−7	1.53e−6
$\Delta x/\Delta B$ (pm/μT)	14.43	7.22	4.81	1.80	1.44	0.90	0.45
$\Delta C/\Delta B$ (aF/μT)	0.29	0.29	0.29	0.29	0.29	0.29	0.29

thickness is required to guarantee mechanical stability and to avoid bending towards fixed electrodes), drawn gap $g = 1.4$ μm$+0.7$ μm, respectively, being the minimum allowed gap by design rule manual and the typical overetch. Provided data are for this specific example but the methodology can be applied using other technologies.

4.2.1 *Sensitivity*

Simulation results for seven different structures are reported in Table 4.1: the only input parameter which is changed is the number of sensing cells N. The purpose of this set of simulations is to design some structures with different number of cells so that the introduced theory about sensitivity can be experimentally proved. As expected from sensitivity equation (Eq. (3.42)), with a correct implementation of the presented theory into the model, mechanical sensitivity does not depend on sensing cells.

The capacitance C_0 and the total mass m_{tot} increase with the number of cells. To a larger capacitance, a larger differential capacitance variation usually corresponds but, in free molecular flow regime, damping is proportional to capacitive area resulting in a flat sensitivity vs. N. Then, if the resonance frequency is kept fixed, the larger mass due to a higher number of cells is counterbalanced by a stiffer structure and so the displacement is reduced. Quality factor has multiple dependencies on both N and spring dimensions according to:

$$Q = \frac{\sqrt{k \cdot m}}{b} = \frac{\sqrt{k \cdot (m_{fixed} + N \cdot m_{cell})}}{2 \cdot b_{area} \cdot N \cdot L_C \cdot H}, \tag{4.3}$$

where m_{fixed} accounts for fixed contribution to the total mass of device and m_{cell} is the mass of each sensing cell and the related portion of central shuttle.

- For a few sensing cells, $N \leq m_{\text{fixed}}/m_{\text{cell}}$ the quality factor has a dependency:

$$Q = \frac{\sqrt{k \cdot m_{\text{fixed}}}}{2 \cdot b_{\text{area}} \cdot N \cdot L_C \cdot H} \propto \frac{1}{N}; \tag{4.4}$$

- For a larger number of cells, $N \geq m_{\text{fixed}}/m_{\text{cell}}$ the quality factor has a dependency:

$$Q = \frac{\sqrt{k \cdot N \cdot m_{\text{cell}}}}{2 \cdot b_{\text{area}} \cdot N \cdot L_C \cdot H} \propto \frac{1}{\sqrt{N}}. \tag{4.5}$$

Simulated values of quality factor are reported and compared to experimental results in Fig. 7.3; results discussion is in Sect. 7.1.2.

4.2.2 *Response Time*

Simulations are performed also to investigate about the effect of MEMS quality factor on system settling time. As already mentioned, to a micromechanical device a second order transfer function with two complex conjugate poles is associated. It can be written as follows:

$$\begin{aligned} T(s) &= \frac{\alpha}{s^2 + \frac{b}{m}s + \frac{k}{m}} \\ &= \frac{\alpha}{s^2 + \frac{\omega_r}{Q}s + \omega_r^2} \\ &= \frac{\alpha}{s^2 + 2\xi\omega_r s + \omega_r^2}, \end{aligned} \tag{4.6}$$

where α is a generic coefficient, $f_r = \omega_r/(2\pi)$ is the mechanical resonance frequency, and the relative damping factor is:

$$\xi = \frac{1}{2Q}. \tag{4.7}$$

From the theory of systems control, it is also known that the envelope of the oscillating response (shown in Fig. 4.3) is an exponential decay with a time constant proportional to the quality factor:

$$\tau = \frac{2Q}{\omega_r} \tag{4.8}$$

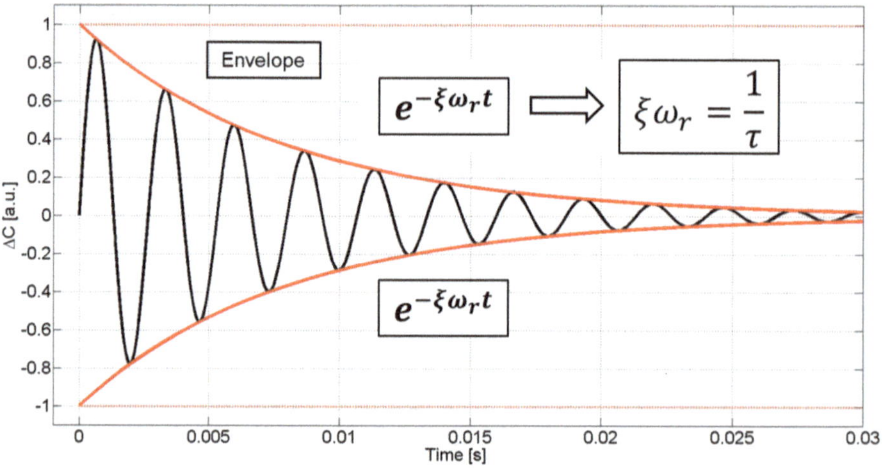

Fig. 4.3 Step response for a second order system with complex conjugate poles: oscillating behavior with exponential decay

On a fixed resonance frequency, the higher the quality factor is, the longer is system response time. According to studies carried out so far, on equal sensitivity, the objective is to implement a smaller device to improve intrinsic noise performance and to integrate it in the same package of other sensors. The resulting quality factor is almost fixed by these constraints, and the remaining parameter to change Q is pressure, depending on manufacturing process. Finally structures with four different values of N (4, 8, 16, and 32) are chosen for implementation. A final remarks must be done on the relationship between Q factor and mechanical bandwidth of devices. Devices with a larger Q take a longer time to reach their maximum oscillating amplitude and this represents a very important aspect considering that this kind of sensors is based on the measurement of capacitive variation amplitude. The increase in time required to get the maximum oscillation amplitude is inversely proportional to device bandwidth and so devices with higher Q can suffer from limited bandwidth. Theoretically there are two possible solutions to this problem: an increase of elastic stiffness and so increase of resonance frequency and, second, a reduction of Q factor. The first approach is not working in the case of this example because resonance frequency is fixed out of audio bandwidth and its increase would worsen the mechanical sensitivity (see Eq. (3.42)). The second approach is more feasible according to packaging limits set by used process.

4.3 Noise Budget Partitioning

This sections deals with the partition of noise budget among different blocks of the system. Starting from high level specifications (i.e., specifications given at the beginning of the project) in terms of maximum allowed noise in the system, critical blocks are identified, defining the specification for each one. Even though this analysis is fundamental as it sets the achievable resolution of the system, the noise budget partitioning is an iterative process of the design. Some design specifications can be defined on the base of preliminary analysis, assumptions, and measures but budget partitioning which will be refined during ASIC design and on the base of consequent experimental results. The required performance for each block is evaluated and in case of any issues the assumptions will be modified. It would be quite difficult at the first run to identify all the critical parameter of the design.

Knowing the minimum magnetic field to be detected and so the maximum level of noise in terms of magnetic field, the overall noise spectral density is obtained. Sum of all noise contributions in the system must not exceed that amount. Supposing to use lock-in-based filtering techniques (this is what is commonly done in many readout schemes for MEMS), a bandwidth of BW = 10 Hz is considered during the following analysis and a minimum detectable field of 1 μT corresponds to an overall equivalent noise spectral density in terms of magnetic field of:

$$\sqrt{S_{B_{\mathrm{TOT}}}} = \frac{B_{\min}}{\sqrt{\mathrm{BW}}} = 316\ \mathrm{nT}/\sqrt{\mathrm{Hz}}\,. \tag{4.9}$$

The two main noise sources are:

1. Brownian thermomechanical noise due to MEMS (in the following referred to as "intrinsic noise");
2. Electronics noise: low-frequency (flicker, RTS), thermal, and quantization noise.

4.3.1 *MEMS Noise*

In the system co-design the noise budget can be equally split between the thermo-mechanical and the electronics noise. This results in a very important specification on the performance in terms of noise of both the MEMS and the ASIC:

$$\sqrt{S_{B_{\mathrm{eln}}}} = \sqrt{S_{B_{\mathrm{mech}}}} = \sqrt{\frac{S_{B_{\mathrm{TOT}}}}{2}} = 223\ \mathrm{nT}/\sqrt{\mathrm{Hz}}\,. \tag{4.10}$$

From the mechanical point of view this level of noise can be obtained, imposing that:

$$\begin{aligned}\sqrt{S_{B_{\mathrm{mech}}}} &= 223\ \mathrm{nT}/\sqrt{\mathrm{Hz}} \\ &= \frac{2k}{Q \cdot I \cdot L} \cdot \sqrt{S_{x_n}} \\ &= \frac{2k}{Q \cdot I \cdot L} \cdot \frac{Q}{k} \cdot \sqrt{S_{F_n}} \\ &= \frac{2k}{Q \cdot I \cdot L} \cdot \frac{Q}{k} \cdot \sqrt{4 \cdot k_{\mathrm{B}} \cdot T \cdot b} \\ &= \frac{2}{I \cdot L} \cdot \sqrt{4 \cdot k_{\mathrm{B}} \cdot T \cdot b}\end{aligned} \tag{4.11}$$

which can be achieved pumping a current into MEMS springs of:

$$I = \frac{2}{L \cdot \sqrt{S_{B_{\mathrm{mech}}}}} \cdot \sqrt{4 \cdot k_{\mathrm{B}} \cdot T \cdot b} \approx 515\ \mu\mathrm{A}_{\mathrm{rms}}\ , \tag{4.12}$$

evaluated for a device with $N = 4$, a spring length of $L = 870\ \mu\mathrm{m}$, result of technology and geometrical constraints and system level optimization; the value of damping coefficient is estimated from theoretical analyses in Sect. 3.3.1. In order to reach the required resolution a current of about 515 $\mu\mathrm{A}_{\mathrm{rms}}$ must be pumped with a consequent intrinsic power consumption of about 500 μW considering a typical MEMS springs resistance of few kΩ.

According to Eq. (4.11), in order to reduce equivalent magnetic field spectral noise density these parameters can be tuned:

- Current flowing through springs with a consequent higher power consumption proportional to the square of current magnification factor (e.g., if the current doubles, the power gets four times higher);
- Length of springs with the compliance of technological constraints about the maximum releasable structure but without paying for a higher power dissipation (on equal elastic stiffness);
- Reduction of damping coefficient and so reduction of package pressure in the range of possible pressures inside a standard industrial package (and staying in the free-molecule flow, hypothesis at the base of these results).

A more detailed discussion about power consumption is reported in Sect. 4.4 and further considerations about the optimization of devices to improve intrinsic noise are given in Sect. 10.1.1.

4.3.2 Electronics Noise

After getting an actual estimation of the required current to reach the level of noise required from the mechanical section of the system, the remaining overall noise budget (a $\sqrt{2}$ lower than the total) is for electronics readout and so:

$$\sqrt{S_{B,\text{eln}}} = 223\ \text{nT}/\sqrt{\text{Hz}}\,. \tag{4.13}$$

Details about electronics noise are given in Chap. 9.

4.4 Power Budget Partitioning

Specifications are oriented to an earth magnetic field detector for consumer applications and so the prototype must be thought to be integrated in portable battery powered devices. Thus, in addition to small dimensions, power consumption is a fundamental aspect in the development of such a system. These are:

$$\begin{aligned} V_{\text{Supply}} &= 1.6\text{–}3.6\ \text{V} \\ I_{\text{MEMS+ASIC}} &= 300\text{–}400\ \mu\text{A}\,, \end{aligned} \tag{4.14}$$

which correspond to a power consumption of 1.4–1.5 mW not to be exceeded. This power budget must be split into the device intrinsic dissipation and the driving and sensing circuitry consumption. For this reason a preliminary analysis about power required by each block becomes interesting.

The first contribution arises from Joule power consumption of Lorentz current flowing through the MEMS springs. According to Eq. (4.11), once the geometry of device and pressure of the package are fixed, the pumped current is the only parameter to act on to improve intrinsic resolution. A trade-off current value must be chosen considering both the intrinsic resolution achieved by the device and the intrinsic power consumption (due to the resistivity of the structural layer forming the springs). Figure 4.4 shows the intrinsic resolution with respect to intrinsic power consumption for a device having $N = 4$ parallel-plate sensing cells.

Two other main parts of the system which have current consumption are driving and readout sections. It is here supposed that driving circuitry consists of an oscillator connected to MEMS driver and some logics. For these circuits (especially for an oscillator) about 50 μW are estimated according to typical power consumptions reported for ring oscillators [3] and relaxation oscillators [4].

The remaining power budget can be used for sensing circuitry. Additional details about an estimation of this contribution are given in Sect. 9.2 considering the used power supply and designed circuits.

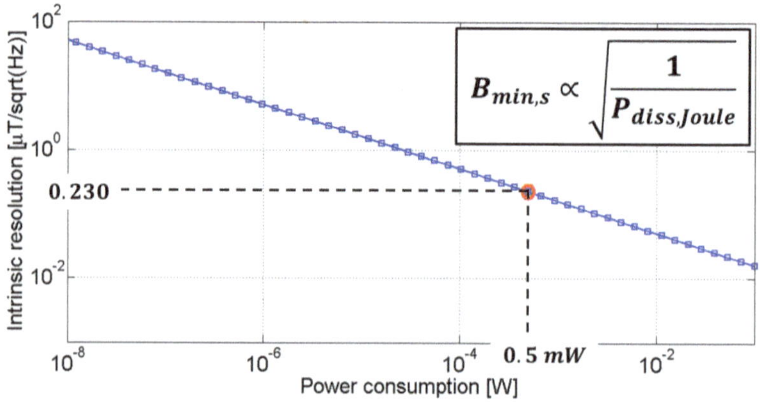

Fig. 4.4 Device intrinsic resolution vs. power consumption due to a Lorentz current flowing through springs

References

1. Z. Izham, M.C.L. Ward, Sens. Actuators, A **115**(2–3), 392 (2004). doi:10.1016/j.sna.2004.04.055. http://www.sciencedirect.com/science/article/pii/S0924424704003358; The 17th European Conference on Solid-State Transducers
2. A. Corigliano, B. De Masi, A. Frangi, C. Comi, A. Villa, M. Marchi, J. Microelectromech. Syst. **13**(2), 200 (2004). doi:10.1109/JMEMS.2003.823221
3. M. Al Kadi Jazairli, D. Flandre, Ph.D. Research in Microelectronics and Electronics, 2009. PRIME 2009 (2009), pp. 312–315
4. U. Denier, IEEE Trans. Circuits Syst. Regul. Pap. **57**(8), 1973 (2010)

Chapter 5
Sensor Design and Fabrication

This chapter deals with devices design and their fabrication. Methodology shown in Sect. 2.1 can be extended with the following steps:

8. Device design;
9. Device mechanical characterization.

Mechanical sensors design and optimization is based on specifications derived from system level simulations. Following steps for prototype development and product development are discussed in Chap. 8.

5.1 FEM Simulations

Simulations using a Finite Element Method (FEM) tool, COMSOL Multiphysics® software, are performed for all implemented devices to double check and refine Simulink results. Particularly, simulations are oriented to check mechanical properties of devices like the requirement of resonance frequency beyond audio bandwidth and elastic stiffness. Among different software available for FEM simulations, COMSOL Multiphysics® provides an interface to draw structures and it solves a set of matrix equations according to geometry, material parameters, and boundary conditions set by the user.

Eigenfrequency analysis is useful to identify device vibrating modes. FEM simulations are performed for each of the four implemented structures. Figure 5.1 shows a 3D simulation for a $N = 4$ magnetometer using a spring width estimated on the base of a typical technology overetch; fixed constraints are set at both springs ends. Results about vibrating modes are reported in Table 5.1. Process height: $H = 15$ μm, spring length $L_s = 870$ μm, actual spring width $W = 5.4$ μm.

C. Buffa, *MEMS Lorentz Force Magnetometers*,
DOI 10.1007/978-3-319-59412-5_5

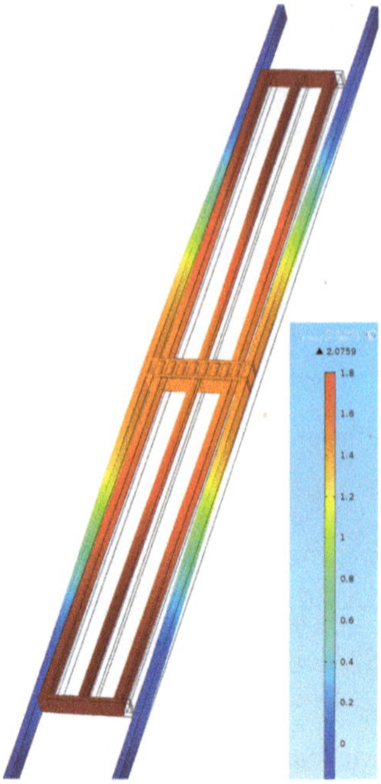

Fig. 5.1 FEM simulations of implemented devices are performed to double check Simulink results

Table 5.1 Vibrating modes for $N = 4$ magnetometer obtained with COMSOL® eigenfrequency simulations

$N = 4$ magnetometer	
I mode	30917 Hz
II mode	70973 Hz
III mode	81676 Hz
IV mode	90113 Hz

A second set of stationary simulations is performed to check devices elastic stiffness. A distributed load is applied on magnetometer springs and simulated displacement is used to estimate elastic stiffness. Results are in accordance with Simulink® model.

5.2 Device Implementations and Layouts

As already introduced in Sect. 3.3.1, devices are fabricated using surface micromachining process ThELMA (Thick Epitaxial Layer for Microgyroscopes and Accelerometers) developed by STMicroelectronics for high-performance and low-cost production of silicon inertial sensors and actuators. ThELMA process allows implementation of suspended structures, using a polysilicon layer of 15 μm height, anchored to substrate through springs. More details about fabrication process are given in Sect. 5.2.1.

Table 5.2 Table reports springs width according to Simulink results, their corresponding values increased by overetch, implemented drawn width taking into account technological constraints and final expected actual width for four different structures

Springs width				
N	4	8	16	32
Simulink actual width	5.39 μm	5.98 μm	6.88 μm	8.12 μm
Theoretical drawn width	6.08 μm	6.68 μm	7.58 μm	8.82 μm
Actual drawn width	6.1 μm	6.7 μm	7 μm	9.1 μm + holes
Expected actual width	5.4 μm	6 μm	6.3 μm	8.4 μm + holes

Four different magnetometers are implemented with different number of sensing cells: 4, 8, 16, and 32. Their layouts are drawn using Virtuoso Layout Suite by Cadence® with the support of ThELMA design rule manual (DRM). For all devices, an overall flexures length of 870 μm is used. Table 5.2 reports drawn springs width and expected actual width according to a total typical overetch of 0.7 μm (each released structure is expected to be thinner of 0.35 μm per side).

For devices $N = 4$ and $N = 8$ drawn springs width are coherent with Simulink® results with an addition of 0.7 μm to take into account polysilicon overetching during process. For device $N = 16$ according to simulated results a drawn width of 7.6 μm is required. This is in contrast with technology design rules according to which a maximum drawn dimension of 7 μm is allowed to guarantee structures release. Thus, the maximum allowed dimension is used for this device spring width (7 μm) expecting an overall lower stiffness and so a lower resonance frequency. Springs width for device $N = 32$ is by far larger to be approximated to a drawn width of 7 μm; in order to cope with the required dimension a 9.1 μm-width spring is implemented with holes (5.3 μm × 3.3 μm each one) to allow chemical etch releasing the structure. The geometry of the springs which can be simulated by the Simulink model is a solid, full bar. The desired spring width for the device with $N = 32$ is larger than the maximum allowed by technology rules and therefore some etching holes must be placed. Since this effect cannot be reproduced in the Simulink model, the equivalence in terms of elastic stiffness between the optimized value from Simulink and the one which can be implemented is verified using FEM simulations.

Figure 5.2 shows layouts of four implemented magnetometers with a detail of *stoppers* which are structures to limit the possible displacement of the moving mass and they must be placed at such a distance that capacitors electrodes cannot get in contact. In this design stoppers are placed in each structure to a distance 200 nm closer than gap and they are connected to a separate pad for biasing. In addition to stopper connection, each device is routed to four pads which are: stator A, stator B, and two pads to access moving mass.

During the full cycle of a product development typically more generations (at least two) of devices are realized in silicon: the first generation, which includes structures above described and shown in Fig. 5.2, is based on theoretical analyses

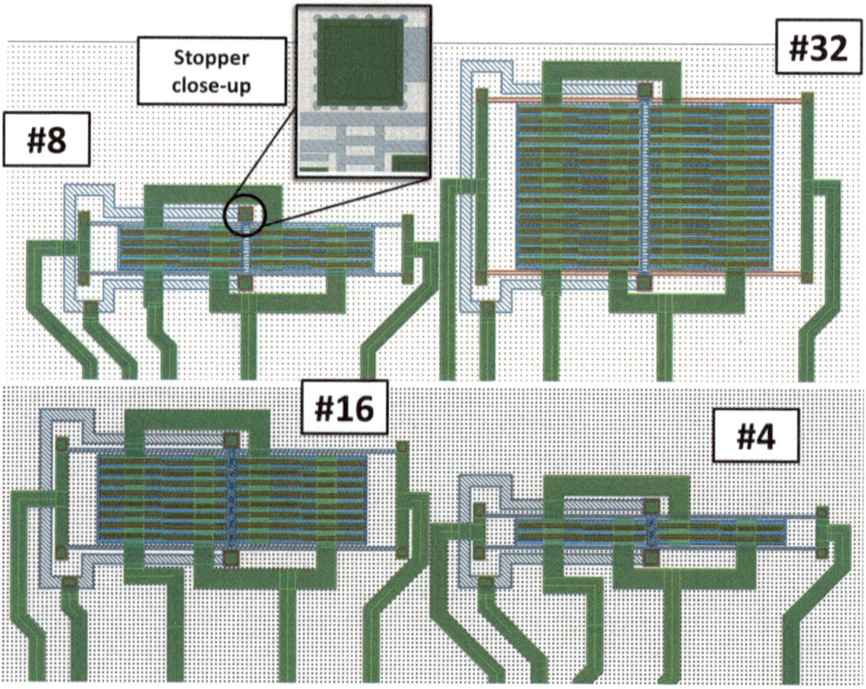

Fig. 5.2 Layouts of MEMS magnetometers

and system design and it is important to validate transduction principle and theory. Subsequently, based on characterization of first devices and after refining model to hardware correlation, a second generation of redesigned structure is implemented with an optimization for mass production. Further details are given in Sect. 10.1.1.

5.2.1 Fabrication Process

Main steps of STMicroelectronics ThELMA process are here summed up and sketched in Fig. 5.3.

1. *Substrate thermal oxidation*: the silicon substrate is covered by a 2.6 μm thick layer of permanent oxide obtained with a thermal treatment at 1100 °C (Fig. 5.3a);
2. *Deposition and patterning of horizontal interconnections*: the first polysilicon layer is deposited above the thermal oxide; this 700 nm thick layer (Thin Polysilicon Layer, shortly TPL) is used to define the buried runners which are normally used to bring electrical signals outside the device (Fig. 5.3b);

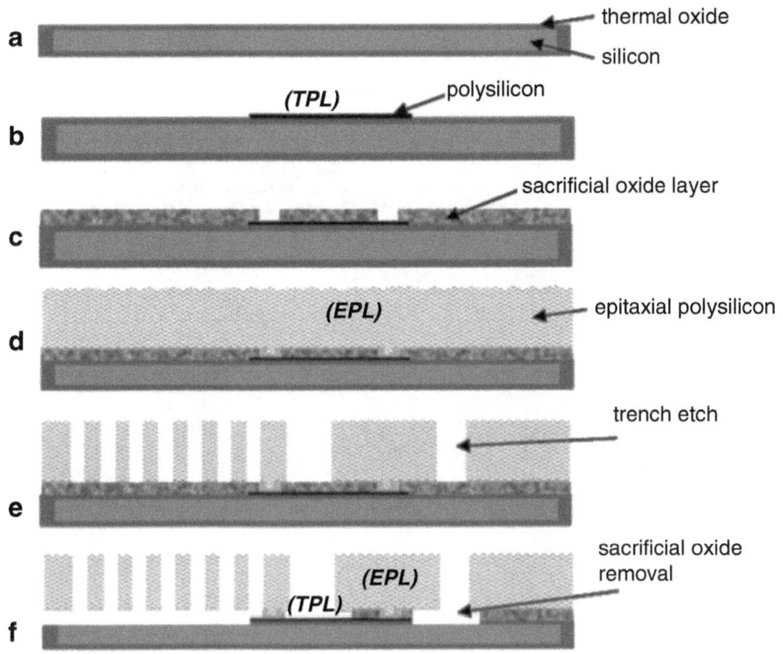

Fig. 5.3 Schematic representation of the different steps of the used STMicroelectronics ThELMA technology, as described in the text

3. *Deposition and patterning of a sacrificial layer*: a thick oxide layer is deposited by means of a Plasma Enhanced Chemical Vapor Deposition (PECVD). This layer, together with the thermal oxide layer, forms a 5.1 μm thick layer which separates the moving part from the substrate (Fig. 5.3c);
4. *Epitaxial growth of the structural layer*: another layer of polysilicon is epitaxially grown in the reactors, reaching the final thickness. In the present work, this layer (so on Epitaxial Polysilicon Layer, EPL) is used to fabricate the inertial mass and its height is 15 μm for the first generation of devices (Fig. 5.3d);
5. *Structural layer patterning by trench etch*: the suspended parts of the mobile structure are obtained through a Deep Reactive Ion Etching (DRIE), with trenches reaching the oxide layer (Fig. 5.3e);
6. *Oxide removal*: the sacrificial and the thermal oxide layer are removed with a chemical reaction (in order to avoid stiction due to attractive capillary forces, this is done in rigorously dry conditions). Typically, the TPL layer is used as a buried runner to provide electrical connections of the structural EPL parts (Fig. 5.3f).
7. *Contact metalization deposition*: the contact metalization is deposited.

The substrate wafer, hosting the released moving structures, is wafer–wafer bonded, through a glass-frit technique, with a suitably patterned and treated silicon capping wafer. This silicon cap is used for different reasons: it constitutes a first

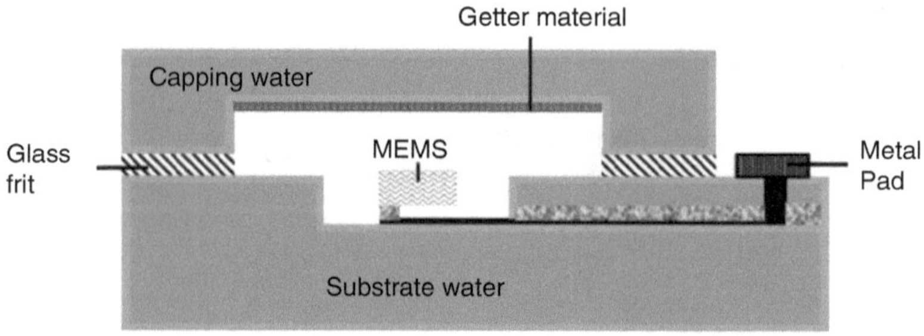

Fig. 5.4 Schematic illustration of the cross-section view of a MEMS packaged device

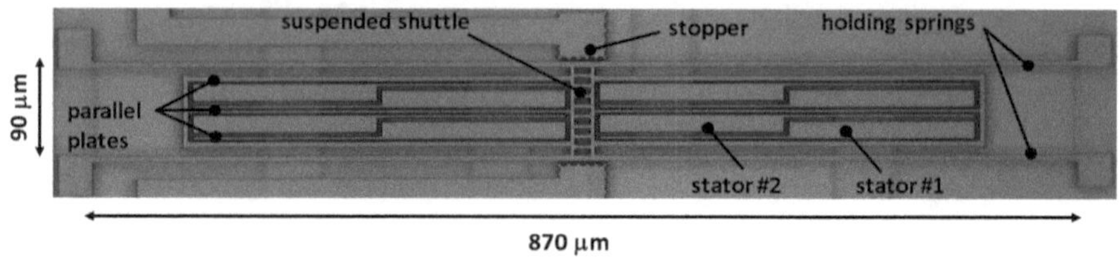

Fig. 5.5 Optical microscope photography of a first generation fabricated device with $N = 4$

protection during the dicing procedure, through which the devices are singled and more important it seals the devices in controlled vacuum environment setting the operating pressure [1]. In particular, for resonant devices it is fundamental to achieve operating pressure around or lower than 1 mbar, in order to achieve reasonably high quality factors. Such low packaging pressures can be obtained by placing getter materials inside the package; these materials are capable of absorbing the molecules degassing at the end of the bonding step [2]. In Fig. 5.4 a cross-section of a MEMS encapsulated in used package is shown.

An optical micrograph of one magnetometer prototype (device with $N = 4$) realized in the first run is shown in Fig. 5.5. This magnetometer has an area of about 870 μm × 90 μm.

References

1. K. Minami, T. Moriuchi, M. Esashi, in *The 8th International Conference on Solid-State Sensors and Actuators, 1995 and Eurosensors IX. Transducers'95*, vol. 1 (1995), pp. 240–243. doi:10.1109/SENSOR.1995.717155
2. B. Lee, S. Seok, K. Chun, J. Micromech. Microeng. **13**(5), 663 (2003). http://stacks.iop.org/0960-1317/13/i=5/a=318

Chapter 6
A Custom Instrument for Electromechanical Characterization

This chapter deals with the development of an instrument for electromechanical characterization of fabricated devices. Precise electromechanical characterization of micromachined structures is of great relevance for several reasons. It represents an experimental validation of theory, behavioral simulations [1] and FEM models [2] of devices; it allows to improve and refine these models through a comparison with experimental data; it enables the estimation of unknown mechanical parameters (for instance, the damping coefficient when packaging conditions are not completely known [3]; or the overetch/underetch with respect to the design layout; or mechanical offsets due to residual stresses); it allows to compute equivalent electrical LCR models for Micro- and Nano-Electromechanical Systems (M/NEMS) resonators [4]; finally, it is a step for testing devices on the industrial scale, to check their yield, reproducibility [5], aging [6], and reliability [7].

Section 6.2 describes working principle, design, and main features of this characterization instrument. This platform is implemented on a PCB with discrete components and then calibrated. Chapter 7 reports electrical measures performed on magnetometers with the developed instrument.

6.1 Motivations for a Custom Instrument

The instrumentation available on the market for electromechanical characterization includes (1) semiconductor parameter analyzers and (2) optical instrumentation for in-plane and/or out-of-plane monitoring of the position of the device suspended structure. The former is basically represented by capacitance meters: these instruments generally have a good (sub-fF) resolution, but they do not allow the dynamic device characterization as their measurement bandwidth is limited. As an example, the *Agilent E4980A Precision LCR meter* [8] provides, at an integration time of $\approx$30 ms (corresponding to $\approx$33 Hz), a maximum fluctuation in the capacitance

C. Buffa, *MEMS Lorentz Force Magnetometers*,
DOI 10.1007/978-3-319-59412-5_6

measurements of 0.5 fF (at 20 mV of test signal amplitude). Similar performance can be obtained using the *QuadTech 7600 Plus Precision LCR Meter* [9]. Optical instrumentation (laser-Doppler and vibrometers) can be used for a full mechanical characterization of MEMS (e.g., the *Polytec PMA-400 Planar Motion Analyzer* [10] has a measurement bandwidth up to 1 MHz). However these instruments require precise optical alignment, stroboscopic cameras, high speed imagers, etc. and are thus very costly (about ten times the electrical ones). Besides, they have a major limit in that they do not allow the analysis of packaged devices: this is a crucial issue, as M/NEMS devices behavior and reliability strongly depend on the packaging conditions[1] (in fact, a measurement at atmosphere pressure does not describe the dynamic behavior of the packaged device). Other examples of research efforts to develop versatile instrumentation for the visual analysis of MEMS devices can be found in [11, 12].

There is no electrical instrument specifically designed for M/NEMS characterization on the market. In this section is presented a prototype of a Mechanical Characterization Platform (MCP), whose aim is filling this lack by providing a complete mechanical characterization for designed magnetometers and, more in general, a lot of capacitive M/NEMS devices. The MCP is intended to be as more versatile as possible because capacitive M/NEMS sensors can differ in terms of process (minimum dimensions, air gap), geometry (characteristic frequency, pull-in voltage), operation (quasi-stationary or resonant), and packaging. The MCP is formed by different elements:

- a differential driving PCB electronic board that is used for capacitive actuation;
- a custom developed, lock-in type, readout PCB electronic board for low-noise and high-sensitivity real-time monitoring of the capacitance variation of the M/NEMS and its conversion into a voltage signal;
- a socket to host either the M/NEMS chip or standard industrial coaxial connectors for wafer-level probe testing;
- a Data Acquisition Board (DAQ) interconnecting the driving/readout PCB to a workstation;
- LabVIEW (by National Instruments™) software libraries for tests automation.

The platform exploits different kinds of electrical stimuli and monitors the resulting capacitance variation, via real-time capacitive sensing. The measurement types are stationary (C-V curves) and dynamic (responses to input steps in the time domain). The data can then be converted into Bode plots in the frequency domain. Experimental results show that the MCP measurement resolution approaches $1\ \text{aF}/\sqrt{\text{Hz}}$, with a temperature drift lower than 100 ppm/°C. The sensing of the capacitance variation is based on a small-amplitude voltage test signal with a frequency of 1 MHz, which restricts the bandwidth range of measurable devices to approximately 100 kHz, as will be detailed in the following section.

[1]Some devices are designed to work at ambient pressure. Nevertheless, in most times a cap on top of device is required to prevent from damages.

6.2 System Overview

The MCP prototype is constituted by a driving/sensing electronic PCB motherboard, a data acquisition board, and software for data processing and displaying. If the wafer is diced, the M/NEMS device can be wire bonded to a standard ceramic carrier or to a small PCB socket that are then connected to the motherboard. Alternatively, the M/NEMS can be on-wafer (wafer is not diced) and a different PCB socket, with coaxial-type connectors, is used for electrical connections between the probe station where the wafer is held and the MCP. A block scheme of the MCP is given in Fig. 6.1. The user selects the input parameters (type of analysis, input waveform, number of averages, etc.) on a LabVIEW interface. Through GPIB connections, the software controls a voltage source that provides the desired stimulus. The software is also connected and synchronized with the readout PCB. More details on the different blocks are given below.

It is now considered a generic M/NEMS capacitive device which includes a differential driving and a differential sensing configuration (see Fig. 6.2), with a suspended structure forming two pairs of capacitances ($C_{d,i}(t)$ and $C_{s,i}(t)$, respectively) with the fixed stators. Capacitances can be of the parallel-plate type (as in the figure), of the comb-finger type or based on fringe effects [13].

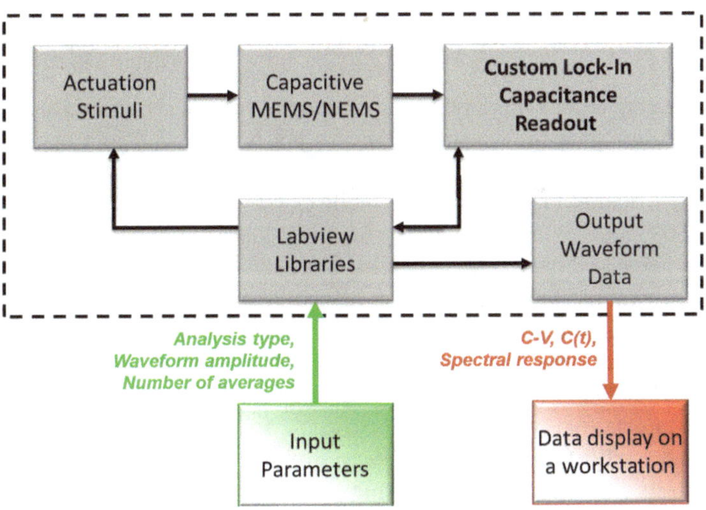

Fig. 6.1 Block-scheme illustration of the Mechanical Characterization Platform. The user can choose the type of measurements and other options (amplitude, averages, initial offset, resolution, etc.) and visualize the results through a LabVIEW interface

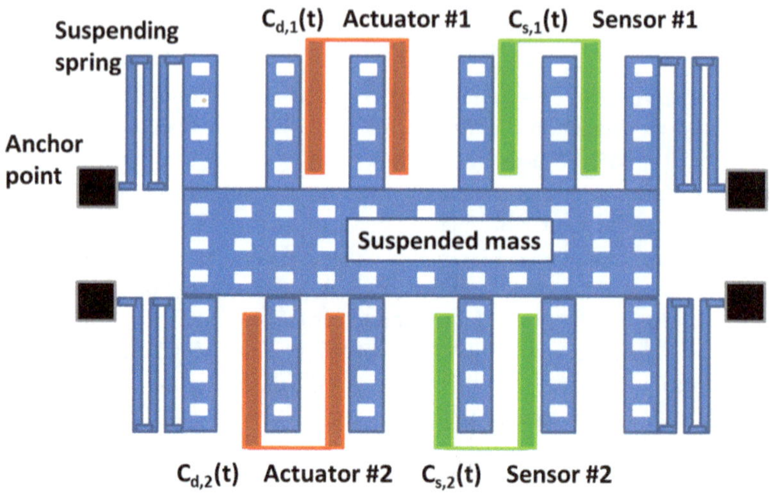

Fig. 6.2 Simplified model of a generic differential-drive, differential-sense, capacitive M/NEMS device

6.2.1 Driving Electronics

Electrostatic driving elements are often part of a M/NEMS device. For some sensors, like gyroscopes, magnetometers or resonant devices [14, 15], driving is always necessary during the device operation. In other situations, like in accelerometers, microphones and pressure sensors driving is not necessary during operation but it may be useful for an initial calibration procedure [16]. For test structures designed to study specific material properties or reliability (e.g. fracture or fatigue tests) driving is mandatory and it is much simpler to embed an electrostatic actuation on-chip rather than using an out-of-chip generated force.

These actuators, often present in the device layout, can be conveniently used to apply electrostatic test stimuli for the device characterization.

In order to make the driving stage of the MCP as more versatile as possible, a differential actuation scheme has been implemented. This scheme gives the possibility to simultaneously apply electrostatic loads in different directions, or to combine the application of both tensile and compressive stresses to a device for reliability tests (the importance of having such a type of actuation is motivated for instance, by the strong dependence of fatigue reliability of polysilicon micro and nano structure on the ratio between the compressive and the tensile stresses during fatigue cycles) [7].

The schematic of the driving stage is represented in Fig. 6.3. Considering a generic periodic input voltage $v_d(t) = A \cdot \mathscr{F}(2\pi \cdot f_d \cdot t)$ provided by an external source (a voltage supply or a waveform generator), where $\mathscr{F}$ is a generic periodic function and A is its amplitude, two output signals are fed to the two M/NEMS driving electrodes:

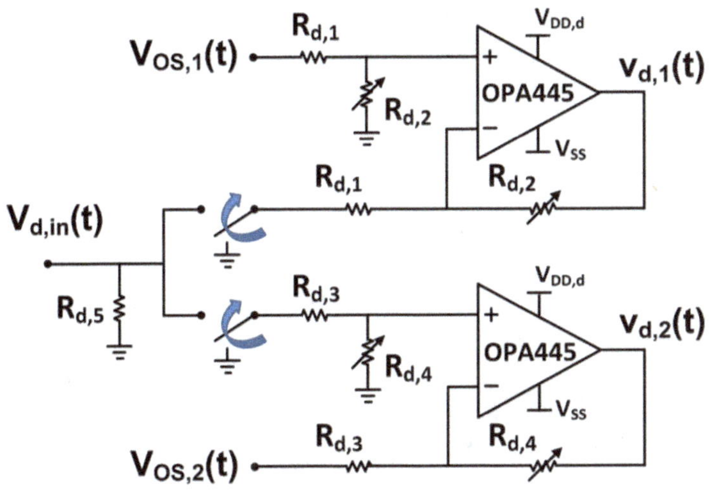

Fig. 6.3 Electronic schematic of the driving stage: two amplifiers generate amplified output signals with a 180° phase difference. The gain and offset of the signals can be selected through the trimming resistors

$$v_{d,1}(t) = V_{OS_1} \cdot \frac{R_{d,2}}{R_{d,1}} + v_{d,in} \cdot \frac{R_{d,2}}{R_{d,1}} \cdot \mathscr{F}(2 \cdot \pi f_d \cdot t + \pi) , \tag{6.1}$$

$$v_{d,2}(t) = -V_{OS_2} \cdot \frac{R_{d,4}}{R_{d,3}} + v_{d,in} \cdot \frac{R_{d,4}}{R_{d,3}} \cdot \mathscr{F}(2 \cdot \pi f_d \cdot t) , \tag{6.2}$$

where $\frac{R_{d,2}}{R_{d,1}}$ and $\frac{R_{d,4}}{R_{d,3}}$ are voltage amplification of the two difference amplifiers and can be set using potentiometers ($R_{d,2}, R_{d,4}$) and V_{OS_i} are offset coefficients, again selectable through trimming resistors (not shown in this simplified schematic). Alternatively, offset voltages can be supplied using an analog output board controlled by LabVIEW software. The driving interface can be clearly used as single ended, either on actuation side 1 or on side 2, by suitably zeroing one of the potentiometers. The selected operational amplifier for these stages is the *OPA445* by *Texas Instruments*, capable of standing up to 90 V of differential voltage supply with a good temperature immunity (typical offset drift ≈25 μV/°C) and a fast slew rate (15 V/μs). Such a large voltage supply range allows the test of very different kinds of devices. A fast slew rate is also required when it is required to apply large step or impulse voltages to the device (these stimuli are not instantaneous as in the ideal situation: yet they should be much faster than the M/NEMS mechanical bandwidth to be measured). Two switches allow to apply an offset voltage only (with no sine stimulus) at one side, while the sine stimulus is still applied at the other side.

The driving stimuli, so far supplied by external instrumentation, can be replaced in future version by monolithic function generators (discrete electronic components that provide waveform generation with the needed accuracy and versatility) or by direct digital synthesizer (DDS).

6.2.2 Capacitive Readout Electronics

The readout electronics developed for the MCP is biased using a ± 5 V voltage source. Its description will be given by highlighting point by point its relevant features:

- *low perturbation of the device*: it is extremely important that during the electromechanical characterization the device behavior is not affected by unwanted electrostatic readout forces. For some capacitive configurations (comb fingers) the effect of the application of a readout electrical signal (a DC or a rapidly varying voltage) is a constant force, independent of the device position. For other configurations (parallel-plate capacitances) the readout force is dependent on the position of the suspended mass and can cause the well-known pull-in instability [17], thus determining the impossibility to test the device. As parallel-plate capacitance sensors have a better sensitivity to area ratio than comb fingers, they are generally adopted for industrial sensors and it thus becomes extremely relevant the development of a low perturbation readout electronics. A further advantage is that, through this approach, also the effect of fringe electrostatic forces, not easily predictable, can be minimized.

  Instead of using a *constant voltage readout* (i.e., with a fixed potential difference applied between the suspended mass contact and the sensing contact) [18], in the proposed MCP the readout signal is obtained from a fast voltage sine wave applied at the suspended mass contact, with each of the sensing electrode kept at virtual ground:

$$v_s(t) = v_0 \cdot \sin(2\pi f_0 \cdot t) \; . \tag{6.3}$$

  The signal is provided by an external waveform generator; in principle, the choice of the amplitude v_0 determines a compromise between the required resolution and the unwanted perturbation (the higher v_0, the larger the sensitivity; the lower v_0, the lower the unwanted readout forces).
- *high sensitivity*: to solve the said compromise and maintain a high sensitivity while keeping a low perturbation, the frequency of the test signal, f_0, must be kept as high as possible. Indeed the current flowing through each sensing capacitance $C_{s,i}$ and into each readout stage, caused by the test signal is:

$$i_{s,i}(t) = C_{s,i} \cdot \frac{dv_s(t)}{dt} + v_s \frac{dC_{s,i}(t)}{dt} \; . \tag{6.4}$$

The second term in the equation above is proportional to the derivative of the capacitance. In quasi-stationary measurements this term is thus null. In dynamic measurements the capacitance change occurs at a maximum frequency equal to the M/NEMS bandwidth f_r. The first term in the equation above is instead proportional to the test frequency f_0. By setting a high $f_0 >> f_r$ a double result is obtained: (1) the current can be simplified as:

$$\begin{aligned} i_{s,i}(t) &= C_{s,i}(t) \cdot \frac{dv_s(t)}{dt} \\ &= C_{s,i}(t) \cdot 2\pi \cdot f_0 \cdot v_0 \cdot \cos(2\pi \cdot f_0 \cdot t) \ , \end{aligned} \tag{6.5}$$

and (2) the current is thus proportional both to the test signal amplitude v_0 and to the test signal frequency f_0, so that the sensitivity can be kept high, even when the amplitude v_0 needs to be low (for the issues described above), by holding a frequency f_0 as high as possible. The same readout principle is adopted in capacitance meters, where the best resolution is obtained at the highest frequency of the stimulus signal.

In the present version of the MCP f_0 can be set up to 1 MHz. This value limits the bandwidth of maximum testable devices to M/NEMS having a bandwidth $f_r < 100$ kHz. The test signal is so far provided by an external waveform generator but it can be replaced by a discrete oscillator, with the required characteristics.

Figure 6.4 shows one channel of the front end electronics, which has now been doubled with respect to the work presented in [19]. This allows a simultaneous differential measurement, an often adopted solution to increase device sensitivity, linearity, and immunity to temperature or other common mode variations. Differential operation is not possible using standard capacitance meters.

The current $i_{s,i}(t)$ of the ith channel flows into the virtual ground of the transimpedance amplifier, ($R_F = 4.75$ kΩ, $C_F = 47$ pF), designed using Analog Devices *OP*42, a fast precision JFET-input operational amplifier with negligible

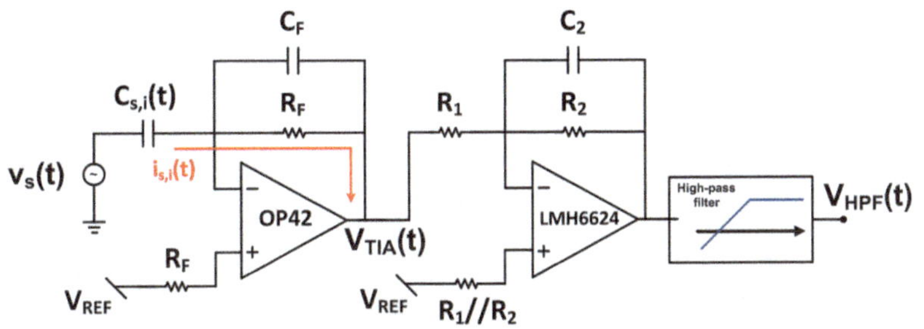

Fig. 6.4 The current flowing in M/NEMS device under test is sensed using a low-noise, low input current and high speed transresistance amplifier. A second gain stage ($G_2 = 47$) is then used to amplify signal to fully exploit demodulation circuit and DAQ board input range. Finally an LC high-pass filter is used to avoid any offset at demodulator inputs

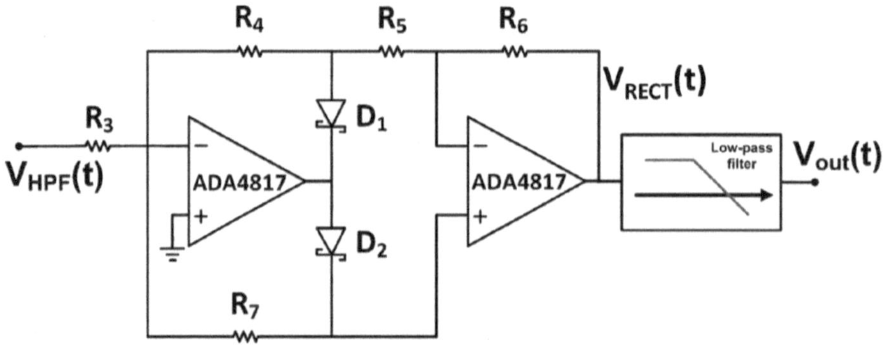

Fig. 6.5 Demodulation is performed through a full-wave diode rectifier: $R_3 = 300\,\Omega$, $R_4 = R_5 = R_6 = 3\,\mathrm{k}\Omega$, and $G = 10$. Fast- and low-noise operational amplifiers combined with Infineon *BAT*62 Schottky diodes are chosen to achieve high frequency demodulation. A closed-loop implementation of the rectifier allows to minimize errors due to turn-on and turn-off diodes time by a factor approximately equal to loop gain

input current noise. The resulting voltage, $v_{TIA}(t)$ is further amplified by a gain stage based on National Semiconductor *LMH*6624. Input bias current of both stages is compensated with resistors placed at non-inverting input of operational amplifiers. A noiseless second order high-pass filter is then used to filter low-frequency noise and offset which can affect the following demodulating circuit.

- *high resolution*: low-noise readout is required as the capacitance variations to be sensed in miniaturized devices are going into the deep sub-fF range. This is accomplished in the MCP through the use of a custom developed lock-in type detection technique. As the information on the capacitance to be measured is carried by a modulated signal in the MHz range, the low-frequency electronic noise (mainly the $1/f$ contribution of the first operational amplifier) can be filtered at this point to improve the signal to noise ratio (SNR). Considering that $1/f$ noise of CMOS based operational amplifiers has a typical corner frequency around 1 kHz, high-pass filtering at 10–100 kHz is a good choice to erase noise and save signals that are centered around 1 MHz.

The schematic of the demodulation stage is shown in Fig. 6.5: after filtering, the signal $v_{HPF}(t)$ is rectified. With respect to previous works by the authors [19], rectifying is now performed using an active diode configuration. Two main advantages of this wholly analog approach are that (1) the need for synchronization signals is eliminated and (2) switching noise and, more in general noise due to digital circuits, is eliminated avoiding cross talk on analog amplifiers. Schottky diodes (D_1 and D_2) are preferred because of their faster turn-on and turn-off times which, combined with a closed-loop solution, are through and through negligible. The rectified output envelope has now the shape of the capacitance variation, so that a low pass filtering at a frequency slightly larger than the M/NEMS resonance frequency gives a demodulated output voltage signal which exactly represents the capacitance variation, in a real-time mode.

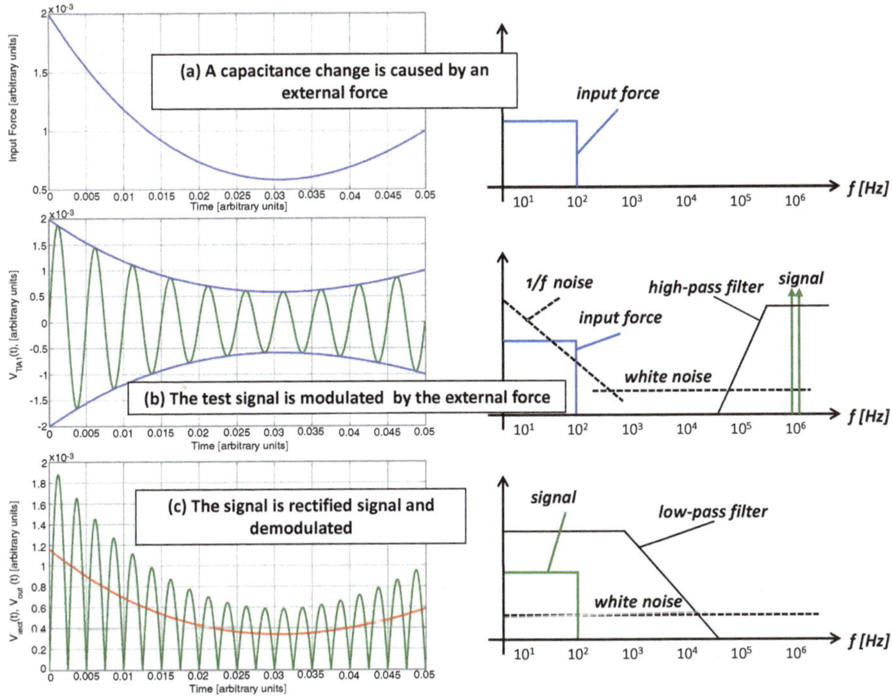

Fig. 6.6 Summary of signals. (**a**) Capacitance rest values change in presence of an acceleration, magnetic field or other forces to be transduced; in the frequency domain input force has a spectrum generally limited to few tens of kHz. (**b**) The application of a high-frequency carrier is fundamental to optimize signals filtering; M/NEMS capacitance variations modulate this high-frequency carriers. (**c**) A low-pass filter (LPF) is required to shift devices signals back to baseband

A set of two different low-pass output filters (selectable by the user through mechanical switches) is provided for each channel.

Figure 6.6b, c summarizes the signals along the MCP readout circuit: in the event of a time varying external force (as in Fig. 6.6a) the sine wave at the output of the transimpedance stage is modulated in amplitude due to the capacitance change, caused by this force. The signal is then rectified and low-pass filtered so that the voltage at the output of the MCP has the same shape as the capacitance change. The representation of the signals in the frequency domain schematically shows the filtering of the low-frequency electronic noise.

6.2.3 Data Acquisition and Display

In order to visualize and process the output data and to control all the measurements from the driving to the display of results, two LabVIEW libraries have been developed. One is provided for stationary, the other for dynamic analyses.

In the stationary C-V characterization the LabVIEW program controls via GPIB a DC voltage source (Agilent E3631A) that is swept between two selectable values. The number of steps and thus their amplitude can be also set by the user through the LabVIEW interface. Each time after the driving voltage is changed, the output voltage of the MCP is sampled through a data acquisition board (DAQ PCI4020) and averaged over a selectable number of samples to further increase the resolution. After that the program sets the new voltage value and the procedure is repeated until the last voltage value is reached. This operation is analogous to the measurement procedure of a capacitance meter. However the operation of such an instrument is limited to quasi-stationary analyses, while the proposed MCP can go beyond.

In the dynamic characterization the device is led to a perturbed position (the corresponding voltage is selected by the user) through a quasi-stationary ramp; then the voltage is instantaneously set to zero. The LabVIEW program, which governs all the measurement, is synchronized with the driving voltage and captures the device capacitance variation after the downward step. Due to the low perturbation of the readout electronics, this corresponds to the device free oscillation. Also in this case the user is allowed to select the number of times the measure is repeated in order to further increase the signal to noise ratio by averaging. Such measurements cannot be done by capacitance meters as their bandwidth is limited. Anyway, the user is free to choose among all standard waveforms or specify arbitrary waveforms according to the proper stimulus required by the device under test.

The user interface of the described acquisitions is reported in Fig. 6.7. For all the measurements the data are also saved as text files, expressed in terms of capacitance variation and applied voltage. By fitting the experimental data or by numerically applying a Fourier Transform to the downward step response, the mechanical parameters of the device and its Bode representation can be extrapolated. This will be shown in detail with an example in Sect. 7.1.1.

Figure 6.8a, b shows the connection to the MCP of wafer-level and single MEMS devices, respectively. Shown in Fig. 6.8c is a differential driving, differential sensing MCP, implemented using a 6-layer technology board whose dimensions are $18.3 \times 8\ \text{cm}^2$.

6.3 Board Characterization: Noise and Sensitivity

The MCP has been characterized using the HP4195A Spectrum Analyzer. The transfer function from the input test signal $v_s(t)$ to the voltage at the output of the high-pass filter $v_{\text{HPF}}(t)$ has been measured by connecting the test signal of

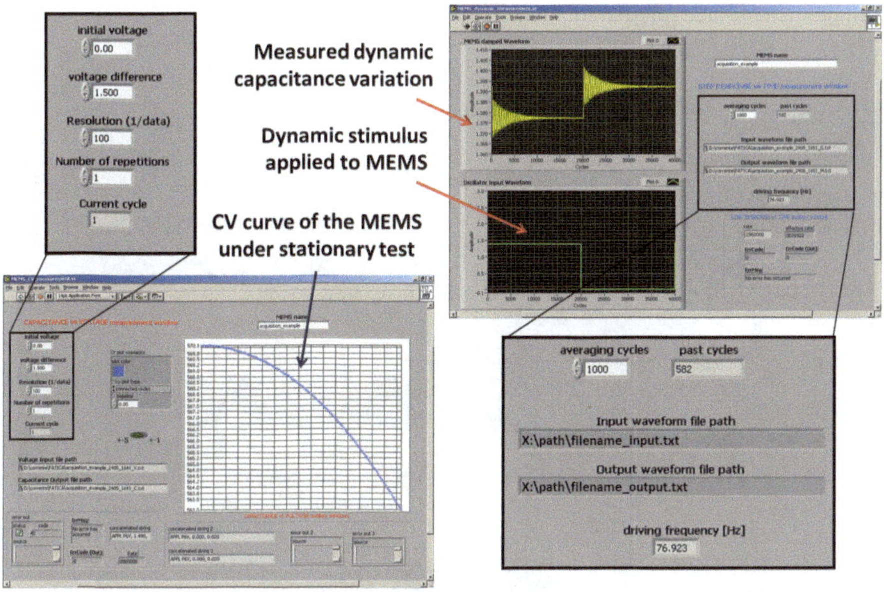

Fig. 6.7 User interface of the static (*left*) and dynamic (*right*) LabVIEW acquisition programs

the analyzer to the input of the MCP, where a 1.02 pF test capacitance (previously measured with the *HP*4274 *LCR* meter) was placed instead of the MEMS sensing capacitance, for correct initial calibration. The test signal had an amplitude $v_0 =$ 450 mV and the frequency swept in the range 10 kHz–10 MHz. Figure 6.9 reports the transfer function measured for the two channels (circle and square markers), compared to the one obtained from *OrCAD Capture* simulations using all the models of the electronic components used in the realized circuit (solid curve). The effect of the high-pass filter in eliminating the low-noise contributions is evident. The transfer function peak occurs around 1 MHz, where the test signal frequency will be conveniently set during the platform operation. The electronic noise was then measured by leaving open the negative input of the transimpedance amplifier while the spectrum analyzer measured the voltage at the output of the MCP, after the low pass filter. For each channel the measurements were repeated for both the low pass filters configurations (10 and 100 kHz). A noise spectral density around 800 $\frac{\text{nV}}{\sqrt{\text{Hz}}}$ is measured (see Fig. 6.10), corresponding to $\approx$1 aF/$\sqrt{\text{Hz}}$ at $v_0 = 450$ mV (measurements are quite disturbed by the proximity to the resolution of the analyzer itself). Using a lower test signal amplitude the unwanted readout force on the device becomes quadratically smaller but, as a drawback, the MCP resolution linearly worsen.

Finally the linearity of the MCP with respect to both the test signal amplitude and the value of the capacitance under measurement was verified, using test capacitances of carefully measured values. First, the parasitic capacitance between

(c)

Fig. 6.8 (**a**) View of the MCP box connected to a probe station for wafer-level measurements. Two power supplies are externally connected. (**b**) Close-up view of the socket for diced MEMS measurements. This socket can be plugged in and out of the motherboard. (**c**) Photograph of a final implementation of prototype motherboard using a 6-layer technology

the two socket plugs, connected to the M/NEMS sensing element and the suspended mass respectively, was evaluated. This parasitic has a value around 435 fF. Other parasitic capacitances to ground do not disturb the measure, as they are kept between fixed voltages (true ground and virtual ground). They do not even afflict the stability of the transimpedance stage as far as their value is lower than about a hundred of pF. The rms value of the filtered output voltage was measured in four different conditions of the input test signal v_0 (100, 200, 500, 900 mV_{PP}), for five different capacitances C_s (195, 259, 468, 1282, 2455 fF—values calibrated using the *HP*4274 LCR meter). Figure 6.11 reports these experimental results, which take

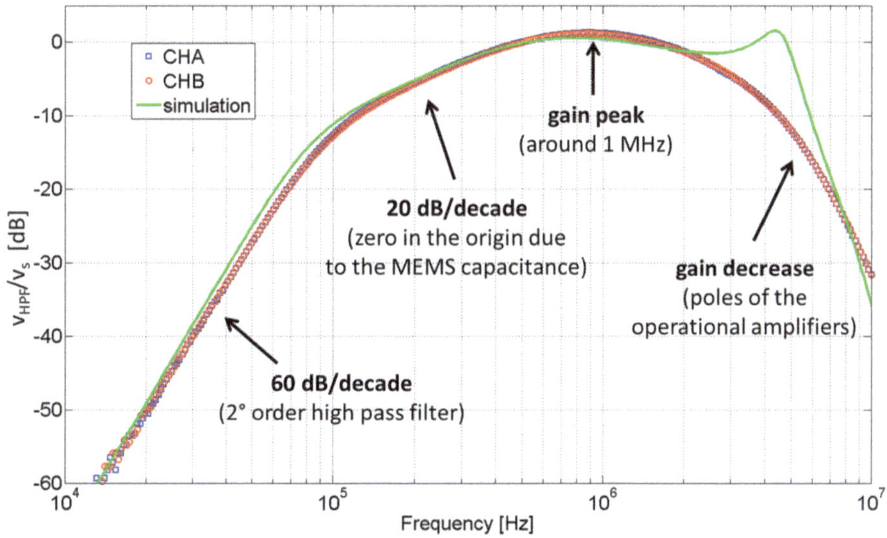

Fig. 6.9 Sensitivity of analog front end of MCP instrument with a test capacitance of 1.02 pF. This transfer function is measured applying a test signal to a test capacitance and reading the resulting voltage at the output of HPF. In the operating region of the MCP (100 kHz and 1 MHz) experimental data fit simulations almost perfectly

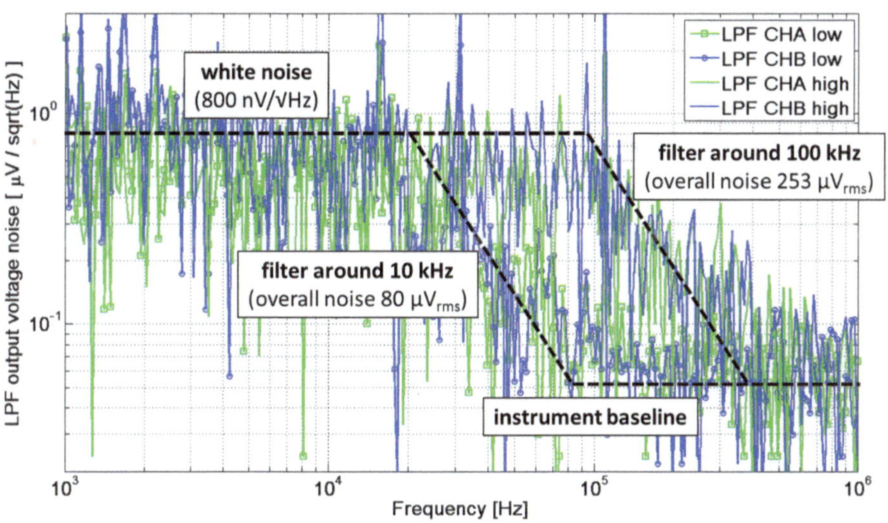

Fig. 6.10 Noise is measured at the output of each channel using HP4195A Spectrum Analyzer

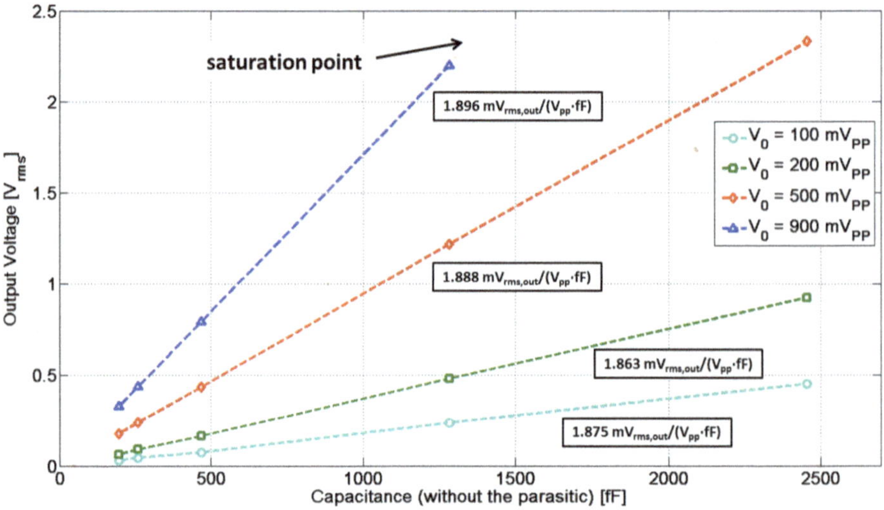

Fig. 6.11 Linearity performance of the presented MCP: measurements are done on a set of four previously calibrated capacitances at four different test signal levels. The evaluated relative linearity error is lower than 2.5%

Table 6.1 MCP performance

Quantity	Symbol	Value
Supply voltage	$\pm V_{DD}$	$\pm$ 5 V
Maximum drive voltage	$v_{a,\max}$	85 V
Test frequency	f_0	1 MHz
Maximum device frequency	$f_{r,\max}$	100 kHz
Resolution	$\sigma_{v,\text{out}}$	$1\ \text{aF}/\sqrt{\text{Hz}}$[a]
Sensitivity	S	1.691 mV/fF[a]
Linearity	ε (%)	< 1
Temperature drift	$\Delta v_{\text{out}}/(v_{\text{out}} \cdot \Delta T)$	< 100 ppm/°C

[a]Measured at $v_0 = 450\,\text{mV}$

into account the presence of the socket parasitic in the conversion between the output voltage and the measured capacitance. The evaluated average sensitivity is:

$$S = \frac{1.879\ \text{mV}_{\text{rms}}}{\text{V}_{\text{pp}} \cdot \text{fF}}\,, \tag{6.6}$$

with a good linearity error which is lower than 2.5%. Table 6.1 summarizes the performance of the MCP.

6.4 Industrialization of the Instrument

In order to transform the presented MCP prototype into a complete product, two blocks need to be further developed: (1) *on-board waveform generators*: both the test signal v_0 and the driving voltage v_d are, in the presented prototype, provided by an external voltage source or waveform generator. The aim is to replace them by using integrated components that allow to generate ramp, sine and square waves with controllable frequency and amplitude; (2) *power supply*: general purpose laboratory power supplies must be replaced by dedicated power supplies.

At the time of publication of this manuscript a product has already been developed with features and specifications reported in [20]. From the technical point of view, as the capacitive M/NEMS developed for scientific and industrial applications have an extremely variable range in terms of rest capacitance, capacitance variation, resonance frequency, air gap, and stiffness (thus pull-in voltage), the versatility of the MCP can be further increased. This can be accomplished through an increase of the maximum measurable motional frequency. It has been shown in Sect. 6.2 how the maximum detectable motion frequency $f_{r,\max}$ is related to the test signal frequency f_0. The improvement thus turns into an increase of the test frequency f_0, which affects and thus involves all the readout chain. Operational amplifier with a larger gain bandwidth product and still a low input-referred noise are required. For what concerns the rectifying stages, diodes in the configuration shown in Fig. 6.5 give a good rectifying up to few tens of MHz. The next-future target is thus to reach 20–50 MHz of maximum test signal frequency with the same readout scheme. Beyond these values, other demodulation schemes would be required (for instance, based on mixers). Note that an increase in the modulation frequency by a factor n also determines an increase in the signal amplitude by the same factor: this means that one can reach a higher SNR, or can keep the same SNR but with a lower electrostatic force applied during readout. This becomes extremely important for nanoelectromechanical systems, where reduced air gap and elastic stiffness can increase the device sensitivity to electrostatic forces and thus make the pull-in voltage very low.

References

1. Z. Izham, M.C.L. Ward, Sens. Actuators, A **115**(2–3), 392 (2004). doi:10.1016/j.sna.2004.04.055. http://www.sciencedirect.com/science/article/pii/S0924424704003358; The 17th European Conference on Solid-State Transducers
2. D. Horsley, in *Handbook of Silicon Based MEMS Materials and Technologies*, chap. 17 (Elsevier, Amsterdam, 2010)
3. G. Langfelder, A. Tocchio, M. Thompson, G. Jaramillo, D. Horsley, in *2010 IEEE Sensors* (2010), pp. 1765–1769. doi:10.1109/ICSENS.2010.5689961
4. C.C. Nguyen, R. Howe, IEEE J. Solid State Circuits **34**(4), 440 (1999). doi:10.1109/4.753677
5. E. Colinet, J. Juillard, L. Nicu, C. Bergaud, IEEE Trans. Instrum. Meas. **54**(4), 1438 (2005). doi:10.1109/TIM.2005.851054

6. J.J. Ruan, N. Monnereau, D. Trémouilles, N. Mauran, F. Coccetti, N. Nolhier, R. Plana, IEEE Trans. Instrum. Meas. **61**(2), 456–461 (2012). doi:10.1109/TIM.2011.2161937
7. G. Langfelder, A. Longoni, F. Zaraga, Sens. Actuators A **159**(2), 233 (2010). doi:10.1016/j.sna.2010.03.011. http://www.sciencedirect.com/science/article/pii/S0924424710001147
8. A. Technologies, Datasheet. http://www.home.agilent.com/agilent/redirector.jspx?action=obs&nid=-34124.3.00&lc=ita&cc=IT&ckey=776077&pubno=5989-4435EN&ltype=LitStation&ctype=AGILENT_EDITORIAL
9. QuadTech, Technical Datasheet. http://www.quadtech.com
10. Polytec Inc., Technical Datasheet. http://www.polytec.com/int/_files/LM_DS_PMA-400_2004_09_E.pdf
11. C. Rembe, R. Kant, R. Muller, in *Proceedings of SPIE*, vol. 4400 (2001), pp. 127–137
12. S. Firebaugh, J. Charles, H.K., R. Edwards, A. Keeney, S. Wilderson, IEEE Trans. Instrum. Meas. **53**(4), 1047 (2004). doi:10.1109/TIM.2004.831504
13. G. Langfelder, A. Longoni, A. Tocchio, E. Lasalandra, IEEE Sensors J. **11**(4), 1069 (2011). doi:10.1109/JSEN.2010.2078499
14. M. Thompson, M. Li, D. Horsley, in *2011 IEEE 24th International Conference on Micro Electro Mechanical Systems (MEMS)* (2011), pp. 593–596. doi:10.1109/MEMSYS.2011.5734494
15. C. Comi, A. Corigliano, G. Langfelder, A. Longoni, A. Tocchio, B. Simoni, J. Microelectromech. Syst. **19**(5), 1140 (2010). doi:10.1109/JMEMS.2010.2067437
16. X. Xiong, Y.L. Wu, W.B. Jone, IEEE Trans. Instrum. Meas. **54**(5), 1739 (2005). doi:10.1109/TIM.2005.855094
17. G. Nielson, G. Barbastathis, J. Microelectromech. Syst. **15**(4), 811 (2006). doi:10.1109/JMEMS.2006.879121
18. M. Lemkin, B. Boser, IEEE J. Solid State Circuits **34**(4), 456 (1999). doi:10.1109/4.753678
19. G. Langfelder, A. Longoni, F. Zaraga, Sens. Actuators, A **148**(2), 401 (2008). doi:10.1016/j.sna.2008.09.011. http://www.sciencedirect.com/science/article/pii/S0924424708004755
20. N. Aresi, *MEMS Characterization Platform*. ITmems S.r.l. Spin-off of Politecnico di Milano (2014). http://www.itmems.it/MCP_rev0.3_web.pdf

Chapter 7
Devices Electromechanical Characterization

This chapter deals with electromechanical characterization of fabricated devices, which validates the proposed theory and used simulation techniques. In general, a complete mechanical characterization is obtained by measuring the stationary and dynamic response of the suspended part of a micro/nano device subject to suitable stimuli. The mechanical behavior of the device can then be described through the derived Bode plots, which give the modulus and phase of the device response with respect to the stimulus frequency. From another point of view, for a single degree of freedom device, the mechanical behavior is that of a mass-spring-damper system (Sect. 3.1.1), which can be described in first approximation through the values of the elastic stiffness k, the damping coefficient b, and the resonance frequency $f_r = \frac{1}{2 \cdot \pi} \cdot \sqrt{k/m}$. For capacitive devices (very popular MEMS on the market in terms of sold parts) the displacement can be indirectly measured through the capacitance variation.

Experimental testing of devices is an important aspect related to MEMS research, to check whether mechanical properties match the design and—in case they do not—to investigate the reasons. Electromechanical characterization of capacitive devices can be done at two different levels: (1) the first one is a quasi-static measurement of capacitance variations with respect to a known external force. This can be enough to evaluate the elastic stiffness in over-damped devices; (2) for resonant sensors instead, only a characterization under dynamic stimuli can be used to evaluate the quality factor Q and the resonance frequency f_r as well.

C. Buffa, *MEMS Lorentz Force Magnetometers*,
DOI 10.1007/978-3-319-59412-5_7

7.1 Time Domain Measurements

7.1.1 Stationary Device Characterization

Quasi-static C-V curves can be used to determine the device elastic stiffness, pull-in voltage, and mechanical offset at rest position. This kind of test is performed in quasi-static regime which means that (1) the driving voltage is applied to the driving electrodes through small increasing steps and (2) after the application of each step, the capacitance is measured after the time required by the mechanical structure to reach a regime condition.

All implemented devices have been characterized according to the shown procedure. For the device characterization, one parallel-plates capacitor set can be used as an actuator (i.e., stator A) and the other set can be used to sense the corresponding capacitance variation (i.e., using stator B). By exchanging the electrodes connections, the measurement can be done in both directions. Measures are performed applying a voltage actuation signal starting from 0 V up to the desired maximum voltage. The implemented LabVIEW software plots capacitance variation with respect to actuation voltage obtaining the so-called quasi-static CV curve.

For each type of device many samples have been tested in order to draw up statistics and test repeatability. All devices under test are in STM standard industrial package with a nominal pressure of 1 mbar. Measures, shown in Fig. 7.1, are performed with a test signal $v_0 = 600$ mV at $f_0 = 1$ MHz.

From CV curves pull-in voltages can be inferred for each structure: for $N = 4$ pull-in voltage varies between 11 and 13 V, for $N = 8$ $V_{pi} \approx 8$ V, for $N = 16$ between 8.5 and 9.2 V and for $N = 32$ $V_{pi} \approx 8.5$ V. Moreover, on each structure measures are repeated after pull-in and CV curves keep the same, which means that for these structures pull-in is nondestructive and mechanical stoppers are efficient. Among all devices, larger structure, $N = 32$ shows a better repeatability justified by the fact that springs width is larger and so process variations less affect elastic stiffness variation and overall mass of the device. From CV curves, elastic stiffness of each device can be inferred and they are (Table 7.1).

Stationary tests show that all devices are correctly released and their capacitance variations correspond to simulated ones with good agreement. Though mechanical residual stresses are present in every structure, the overall symmetry is good in both directions. As mentioned before, each device has been tested applying a driving signal first on stator A and then on stator B. The shown C-V results demonstrate that, with respect to the completely symmetric design, an asymmetry is present on the device. This asymmetry can be ascribed to the layout of the suspending springs, which suffer from residual thermal stresses during the process. From a process optimization point of view, a statistical analysis on such asymmetries could be used to estimate the intensity of thermal stresses. From the designer point of view, the result gives guidelines for a more reliable and repeatable device design.

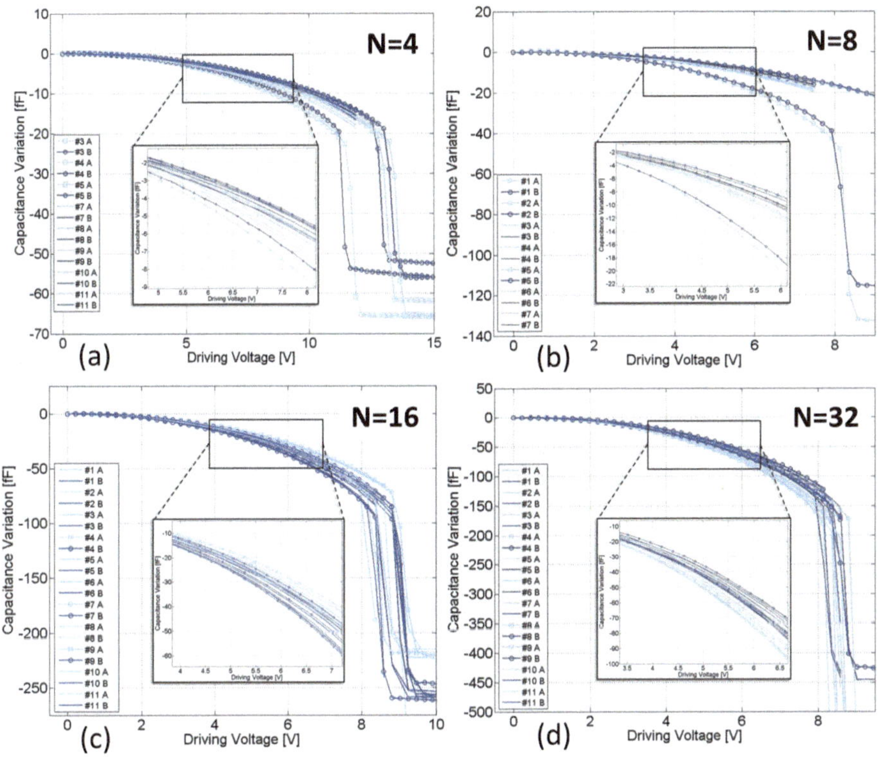

Fig. 7.1 Stationary measures: capacitance variations between moving mass and sensing stator with respect to applied driving force for each structure: (**a**) $N = 4$, (**b**) $N = 8$, (**c**) $N = 16$, (**d**) $N = 32$. *Insets* show measures spreads before pull-in: the lower spread for $N = 32$

Table 7.1 Devices elastic stiffness estimation based on CV experimental results

Elastic stiffness		
N	k (Measured) (N/m)	k (Simulink) (N/m)
4	17.85	17.6
8	24.50	24.1
16	28.35	36.8
32	57.05	60.5

For comparison, Simulink model values are reported. All predicted values are in good agreement with measured ones; device $N = 16$ mismatch is justified by its layout implementation, as detailed in Sect. 5.2

7.1.2 Damping Measure and Experimental Proof of Presented Theory

Dynamic test is a fundamental step in electromechanical characterization of devices to evaluate resonance frequencies and quality factors and so to derive damping coefficient. The first two parameters can directly be measured and, with a good estimation of the mass through available models, damping coefficient can be estimated on the base of experimetal values.

For the device dynamic characterization, a higher frequency stimulus is applied to the device under test. In particular, a square waveform or a saw-tooth waveform is applied to the driving electrode of the device. During the first semi period, the device is led to a perturbed position; during the second semi period, the device is released toward the rest position. In case of under-damped devices, the free mechanical oscillation and the quality factor can be measured in this latter time interval. In case of over-damped devices, the transient time toward the rest position can be used to evaluate the device mechanical bandwidth.

In some applications, like in Lorentz force-based magnetometers, the evaluation of the resonance frequency and the quality factor of the structure are very important because the amplification obtained when operating the device at the resonance frequency can be exploited to enhance the device sensitivity as previously detailed in Sect. 3.2.

As mentioned in Sect. 4.2.2, a MEMS is a second order system with two complex conjugate poles whose frequency response shows an overshoot which depends on damping coefficient of the system related to quality factor according to:

$$Q = \frac{1}{2\xi} \,. \tag{7.1}$$

Higher is Q, higher is maximum percentage overshoot with respect to rest position which can be estimated:

$$S_{\%} = 100 \cdot \exp\left(\frac{-\pi\xi}{\sqrt{1-\xi^2}}\right) = 100 \cdot \exp\left(\frac{-\pi}{\sqrt{4Q^2-1}}\right) \,. \tag{7.2}$$

For this reason the applied stimulus amplitude must be low enough to avoid pull-in because of overshoot. Sensor response to an upward or downward step of a driving square wave is a decaying sinusoid with an exponential envelope according to:

$$\frac{C(t)}{C_\infty} = \left(1 - \frac{e^{-j\omega_r \cdot t}}{\sqrt{1-\xi^2}} \cdot \sin(\omega_r \cdot t \cdot \sqrt{1-\xi^2} + \arccos\xi)\right) \,. \tag{7.3}$$

Experimental resonance frequency can be inferred measuring the time between two consequent peaks in the oscillation:

$$f_r = \frac{\omega_r}{2\pi} = \frac{1}{T} \,. \tag{7.4}$$

Envelope decay time constant is:

$$\tau = \frac{1}{\xi \cdot \omega_r} = \frac{2Q}{\omega_r} \,. \tag{7.5}$$

Figure 7.2 shows measured free oscillation of the described magnetometers and their experimental mathematical fitting obtained with a custom Matlab program. For device $N = 4$ a resonance frequency of $f_{r,4} = 28.3$ kHz is measured, for $N = 8$ $f_{r,8} = 28$ kHz, for $N = 16$ $f_{r,8} = 26.2$ kHz, and for $N = 32$ $f_{r,32} = 27.8$ kHz. Measures values are comparable to designed resonance frequency value of $f_{r,\text{nom}} = 28.3$ kHz and variations can be justified by process variations. Magnetometer with $N = 16$ is a specific case, as already explained: its springs width is thinner than designed one because of the maximum releasable dimension in the fabrication process. Designed value should have been $W_{\text{spring},8} = 7.6$ µm but maximum releasable dimension is $W_{\text{release}} = 7$ µm; thus difference in resonance frequencies is justified.

For every kind of structure, measurements have been repeated for three different samples. Figure 7.3 summarizes the results, showing through circle markers the obtained quality factor as a function of the number N of differential parallel-plate cells. The light dotted curve is the theoretical prediction from Simulink® (including the damping model) described in Sect. 4.1, at the nominal resonance frequency and the nominal pressure. The dashed curve with cross markers is the result of the same simulation, at the best fitting pressure, which turns out to be a factor $\approx$1.3 higher than expected ($b_{\text{area}} = 7.54\ \text{kg}/(\text{s}\,\text{m}^2)$).

The results confirm the theoretical modelling, showing an initial inverse dependency of the quality factor on the number N of differential sensing cells, justified by the proportionality to N of the damping coefficient b:

$$Q = \frac{2 \cdot \pi \cdot f_r \cdot m}{b} = \frac{2 \cdot \pi \cdot f_r \cdot (m_{\text{fixed}} + N \cdot m_{\text{cell}})}{2b_{\text{area}} \cdot N \cdot L_C \cdot H} \,, \tag{7.6}$$

m_{fixed} being the mass of the shuttle (which can be neglected only for $N >> 1$), and m_{cell} being the mass associated to every added sensing cell in the structure. Means to electronically enhance the quality factor have been suggested, for instance, in [1], but are not considered here as they require further circuitry at the cost of an increased power dissipation.

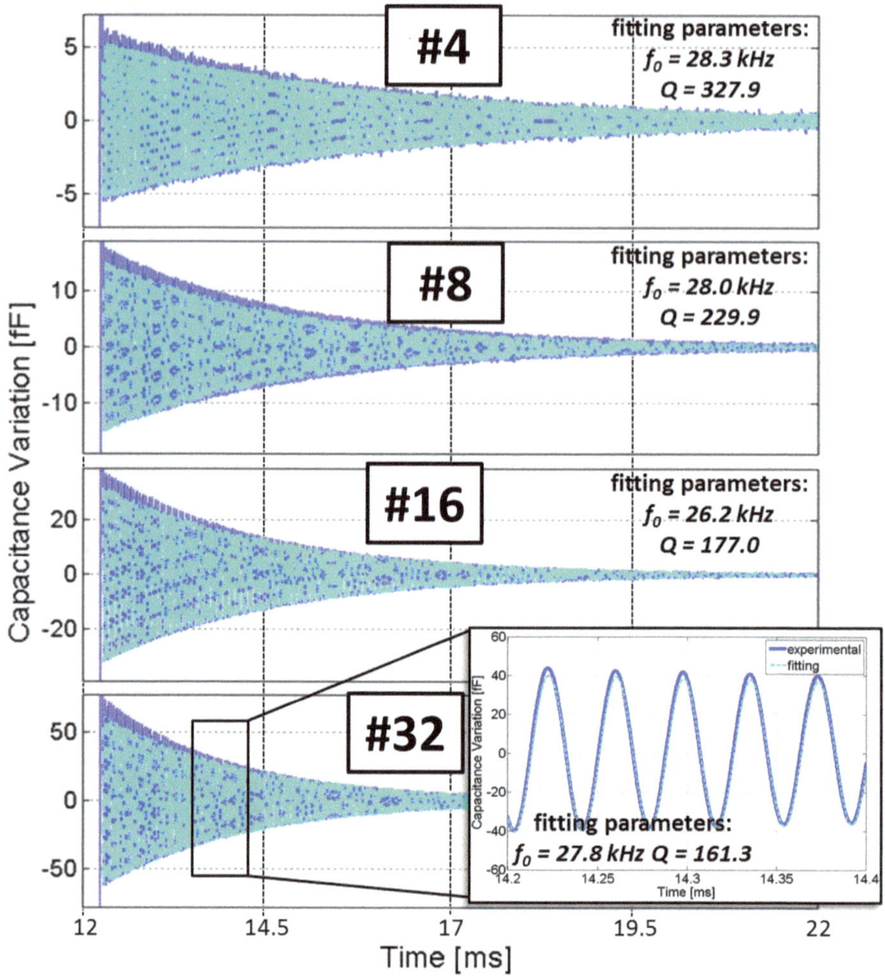

Fig. 7.2 Capacitance variation after a downward step response for the four different devices tested in this work. The experimental curves are superimposed to the best fitting from which the resonance frequency and quality factor can be estimated

7.2 Spectral Responses

Section 7.1 deals with electromechanical characterization in the time domain exciting the device under test both at a very low frequency (DC or quasi-static) and at resonance. A frequency analysis allows to study devices behavior in a more extended range of frequencies and to validate the resonator model implemented in Simulink® software. The experimental setup is shown in Fig. 7.4; it consists of HP4195A Spectrum Analyzer to apply an excitation voltage to a stator and measure

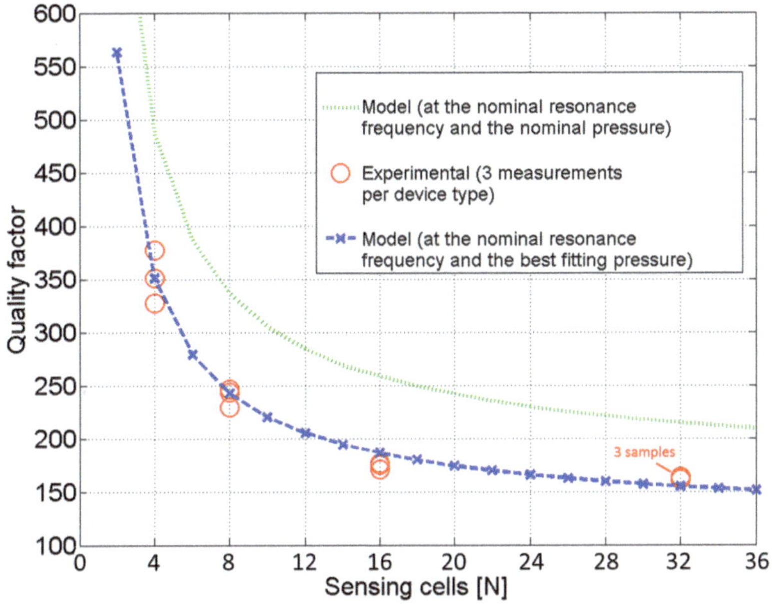

Fig. 7.3 Theoretical model for the quality factor of the four different devices tested in this work, at the nominal pressure of 1 mbar (*light curve*). The experimental results (*circle markers*) fit with a curve corresponding to a packaging pressure 1.35 times larger (*dashed curve with cross markers*)

the transfer function sensing the voltage at the output of the transimpedance amplifier. Moving mass is biased at a voltage V_{MM}. The sensing electrode (represented by the other stator in the differential sensing cell) is kept at virtual ground through the feedback amplifier.

Device transfer functions are evaluated for different biasing voltages V_{MM} for an instrument output power of −30 dBm corresponding to 10 mV_P on the stator. By analyzing magnetometers spectral responses it is possible to evaluate resonance frequencies under different biasing condition. In Fig. 7.5 measures performed on four different devices are shown: increasing the biasing voltage on moving mass electrode, peak amplitudes increases and resonance frequency shifts to a lower value. This frequency shift as a function of the biasing voltage is justified by the effect of an electrical stiffness which is directly proportional to the square of voltage:

$$k_{eq} = k_{mec} + k_{el} = k_{mec} - \frac{A_C \cdot \varepsilon_0 \cdot V^2}{g^3}, \tag{7.7}$$

$$f_r = \frac{1}{2\pi} \cdot \sqrt{\frac{k_{mec} + k_{el}}{m}}. \tag{7.8}$$

Table 7.2 reports, for each structure, measured minimum and maximum resonance frequency for a moving mass biasing between 1 and 10 V.

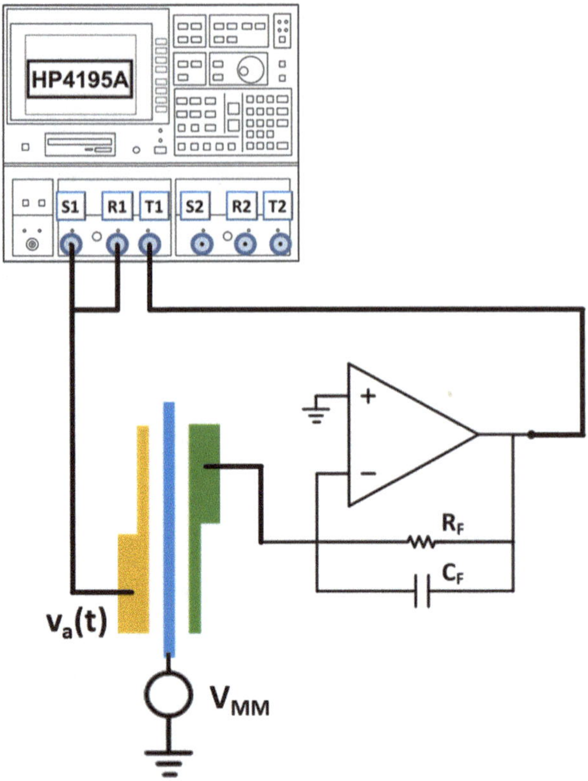

Fig. 7.4 Schematic illustration of the setup used to characterize frequency response of magnetometers: HP4105A Spectrum Analyzer is used to measure device transfer function

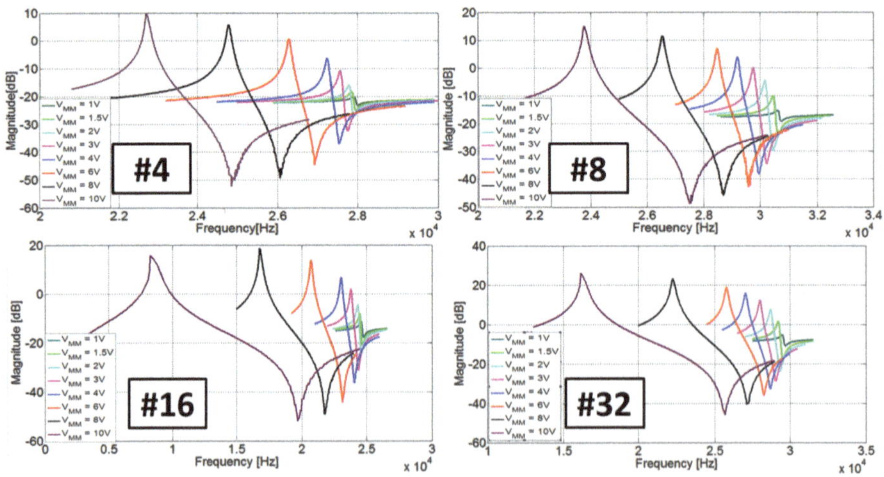

Fig. 7.5 Spectral response magnitude of the four implemented magnetometers for different values of V_{MM}

Table 7.2 Minimum and maximum resonance frequency to due electrostatic spring softening for the four implemented structures

Struttura	$f_{r,\max}$ @ $V_{MM} = 1$ V (kHz)	$f_{r,\min}$ @ $V_{MM} = 10$ V (kHz)
Struttura $N = 4$	27.92	22.7
Struttura $N = 8$	30.63	23.77
Struttura $N = 16$	24.6	8.35
Struttura $N = 32$	29.52	16.24

The current measured at sensing electrode (electrode connected to transimpedance amplifier) should only have a component due to capacitance variation. The presence of an antiresonance peak in the measures can be explained by the presence of a second contribution of current flowing in sensing electrode which is due to a feedthrough capacitance between the driving and the sensing electrode. More details about effects of a feedthrough capacitance can be found in [2].

References

1. M. Thompson, D. Horsley, in *Solid-State Sensors, Actuators and Microsystems Conference, 2009. TRANSDUCERS 2009. International* (2009), pp. 1194–1197
2. A. Tocchio, New paths towards low-power low-noise MEMS accelerometers. Ph.D. thesis, Politecnico di Milano (2012)

Chapter 8
Driving and Readout Electronics: A Discrete Components Solution

Referring to methodology introduced in Sect. 2.1 the next proposed steps are:

10 Development of an experimental setup for magnetic field measurements: after mechanical characterization of devices, which has the purpose to verify that parameters of fabricated sensors fulfill specifications, a setup for measurements in presence of an external magnetic field must be built.
11 Design of driving and readout electronics with discrete components to validate the prototype.

8.1 Driving Electronics

With respect to some other micromachined sensors, like accelerometers, pressure sensors, microphones, etc., Lorentz force magnetometers require a driving block to pump a current flowing into springs. Their sensitivity, intended as capacitance variation per magnetic field variation (see Sect. 3.3), depends on the amplitude of the AC current used for excitation. This current can be provided by either applying a voltage over the flexures or directly forcing a current into them.

The first approach can be implemented with a simpler circuitry (i.e., a voltage oscillator can be connected to the ends of springs) representing a costs and area effective solution for final product implementation and mass production where power consumption and occupied area play an important role. On the other hand, using a voltage approach, uncertainty of flowing current depends on uncertainty of springs resistance which varies because of the manufacturing process. This uncertainty about the actual flowing current affects system sensitivity from unit to unit. This issue could be overcome with the regular calibration that all manufactured units are subject to before being sold on the market.

C. Buffa, *MEMS Lorentz Force Magnetometers*,
DOI 10.1007/978-3-319-59412-5_8

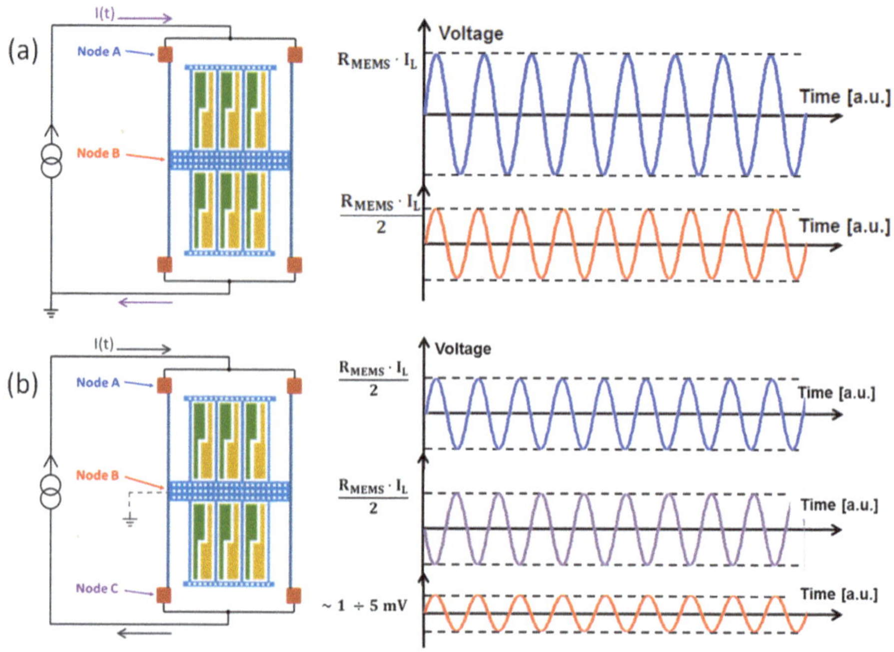

Fig. 8.1 Unbalanced (**a**) and balanced (**b**) scheme for magnetometers current excitation. Using configuration shown in (**b**), the residual signal at moving electrodes can be minimized with a relevant reduction of unwanted electrostatic forces

The latter approach, based on forcing a current into the structure, is instead more accurate and independent on process variations, flexures resistance, on-chip, and external connections. The main drawback of this approach is the requirement of a more complex solution to implement a circuit which behaves as a current generator. For device accurate characterization a current solution is preferred with the main advantage for the user of being able to control the precise value of the current pumped into device springs.

Another fundamental aspect to be considered in development of a driving circuit is related to electrostatic forces arising from the current flowing in the resistive flexures. If the moving mass is connected between the current generator and a reference (GND), as shown in Fig. 8.1a, the potential of central shuttle and so the potential of rotors depends on both amplitude and frequency of pumped current. In case of an AC driving at resonance to exploit Q amplification, rotors voltage is an AC signal whose amplitude is $R_{\text{MEMS}} \cdot I_L/2$ (supposing that process mismatch does not affect springs symmetry). Ideally, this voltage signal on rotors generates two equal electrostatic attractions (towards stator A and stator B) and the moving mass does not move. Nevertheless, because of overetch spread and residual stresses the moving mass might have a mechanical offset. Thus electrostatic forces, which are modulated at the resonance frequency by driving signal, are slightly different and moving mass can displace, generating a false output signal even without an external magnetic field. In order to minimize these forces, a balanced fully differential solution has been studied according to what was previously proposed by Kyynäräinen et al. [1]. Middle point of the springs is now kept to a constant voltage: on one end of the spring the voltage increases while on the other end decreases, ideally, of the same value (the two ends basically have an anti-phase voltage behavior). There might actually be a residual signal at f_r on moving electrodes because of springs resistance mismatch and driving circuit nonideal behavior. Yet, this residual signal is much lower than an unbalanced solution. Figure 8.1b schematically represents signals at MEMS springs ends with a balanced driving.

Different circuit topologies to implement a current generator are analyzed including circuits based on *JFETs* and closed-loop circuits based on operational amplifiers and *BJTs*. Their need for transistors biasing does not make them a suitable solution for an AC sink and source driving without a DC current level. On the contrary, the Howland current pump [2], another current generator scheme used in applications where a high accuracy in current generation is required [3], satisfies the requirements for excitation:

- Bandwidth: up to MEMS resonance frequency;
- No DC current offset (in presence of a magnetic field, a DC current component gives a displacement with respect to rest position which depends on the field itself);
- Bidirectional: sink and source;
- Linear range of about 1 μA–10 mA to test devices with low and high magnetic fields and varying the current.

The chosen implemented scheme for MEMS driving consists of a bipolar (current can be either sunk or sourced) current generator based on an improved Howland current pump and a circuit to make the driving differential. The next subsections introduce the operating principle of Howland current generator and its main features, then design and implementation are shown.

8.1.1 Improved Howland Current Pump

Figure 8.2 reports the schematic of an improved Howland current pump, which is an operational amplifier topology which forms a linear voltage to current conversion providing very high output resistance [2]. Input signal is a voltage stimulus, either DC or AC or a combination of both and output signal is a current proportional to input voltage through a resistor. The main advantage of this circuit over other current topologies is that it does not make use of transistors or switches and, in case of dual power supply, it produces both positive and negative currents without distortion close to zero level. A main drawback is the worsening of output impedance (and other parameters as discussed in the following) if resistors are not well matched.

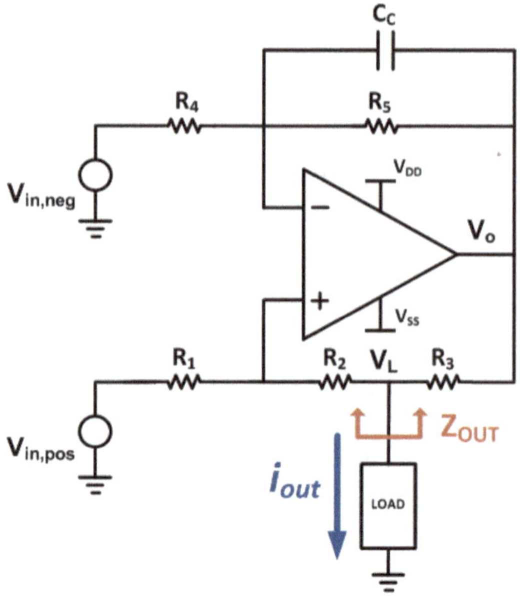

Fig. 8.2 Driving circuit consists of a voltage to current converter based on an improved Howland current pump *(a)* and a feedback circuit to balance MEMS and keep rotors potential close to ground

Before proceeding to a detailed description of implemented design, a general description of the circuit is shown in order to set some preliminary constraints for a correct operation and to calculate output current vs. input voltage transfer function. Referring to Fig. 8.2:

$$\begin{aligned} V^+ &= \frac{R_1}{R_1 + R_2} \cdot \left(V_L - V_{\text{in,pos}}\right) + V_{\text{in,pos}} \\ &= \frac{R_1}{R_1 + R_2} \cdot V_L + \frac{R_2}{R_1 + R_2} \cdot V_{\text{in,pos}} \, . \end{aligned} \tag{8.1}$$

Voltage at operational amplifier output is:

$$\begin{aligned} V_o &= \left(1 + \frac{R_5}{R_4}\right) \cdot V^+ \\ &= \left(1 + \frac{R_5}{R_4}\right) \cdot \frac{R_1}{R_1 + R_2} \cdot V_L \\ &\quad + \left(1 + \frac{R_5}{R_4}\right) \cdot \frac{R_2}{R_1 + R_2} \cdot V_{\text{in,pos}} \, . \end{aligned} \tag{8.2}$$

The output current is:

$$
\begin{aligned}
I_{\mathrm{out}_L} &= \frac{V_{\mathrm{in_{pos}}} - V_L}{R_1 + R_2} + \frac{V_o - V_L}{R_3} \\
&= \frac{V_{\mathrm{in_{pos}}} - V_L}{R_1 + R_2} + \frac{V_o}{R_3} - \left(\frac{1}{R_1 + R_2} + \frac{1}{R_3}\right) \cdot V_L \\
&= \left(\frac{1}{R_1 + R_2} + \frac{1}{R_3} \cdot \frac{R_4 + R_5}{R_4} \cdot \frac{R_2}{R_1 + R_2}\right) \cdot V_{\mathrm{in_{pos}}} \\
&+ \left(\frac{1}{R_3} \cdot \frac{R_4 + R_5}{R_4} \cdot \frac{R_2}{R_1 + R_2} - \frac{R_1 + R_2 + R_3}{(R_1 + R_2) \cdot R_3}\right) \cdot V_L
\end{aligned}
\tag{8.3}
$$

In order to guarantee that this circuit operates as close as possible to an ideal current generator, the pumped current must not depend on the load and so second term in Eq. (8.3) must be null for any value of V_L; thus:

$$
\frac{1}{R_3} \cdot \frac{R_4 + R_5}{R_4} \cdot \frac{R_1}{R_1 + R_2} = \frac{R_1 + R_2 + R_3}{(R_1 + R_2) \cdot R_3} \tag{8.4}
$$

resulting in the condition:

$$
\frac{R_4}{R_5} = \frac{R_2 + R_3}{R_1} \tag{8.5}
$$

Eq. (8.5) must be satisfied for a correct operation of the circuit and, without a resistive load, positive and negative loops tend to be equal (details will be given in the following). Output current vs. input voltage transfer function can be obtained from Eq. (8.3) with some reductions:

$$
I_{\mathrm{out},L} = +\frac{R_5}{R_4} \cdot \frac{1}{R_3} \cdot V_{\mathrm{in_{pos}}} \tag{8.6}
$$

Under the same hypothesis and considering Eq. (8.5) a similar result is obtained for negative input:

$$
I_{\mathrm{out},L} = -\frac{R_5}{R_4} \cdot \frac{1}{R_3} \cdot V_{\mathrm{in_{neg}}} \tag{8.7}
$$

On the whole, the circuit pumps an output current which is:

$$
I_{\mathrm{out},L} = \frac{R_5}{R_4} \cdot \frac{1}{R_3} \cdot (V_{\mathrm{in_{pos}}} - V_{\mathrm{in_{neg}}}) \tag{8.8}
$$

Eq. (8.8) shows that output current in an improved Howland current pump is proportional to both voltage inputs through some resistors. Choosing $R_4 = R_5$ the relationship can further be reduced so that the current only depends on one resistor with the consequent better accuracy.

Moreover, it can be demonstrated that output resistance:

$$R_{\mathrm{out_{How}}} = \frac{R_1 + R_2}{R_2 + R_3} \cdot \frac{R_3}{\pm \mathrm{tol}} \tag{8.9}$$

is directly related to the value of input resistors and to resistors tolerance or resistor mismatches.

Effects related to operational amplifier input bias current, offset, output swing, and slew rate are reported in [4]. Additional remarks about output impedance and stability analyses (evaluation of loop gain of the circuit) with related compensation techniques can be found in [5]. And, finally, a detailed noise analyses of a Howland current source is presented by Hammond et al. [6]. Calculations and equations are not reported as the reference literature is comprehensive.

8.1.1.1 Components Choice

This circuit is designed to pump an output current up to 10 mA (this is not the level of current allowed for normal operation because of a too high power dissipation, however the circuit must be versatile for different kind of tests) into typical MEMS resistances up to about few kΩ (relative high voltage arises at MEMS springs ends). Since output current is supplied by operational amplifier, a first specification on the choice of active element is set. Different commercial operational amplifiers are considered for implementation and *OPA*627 by Texas Instruments turns out to be a good choice because it can be powered up to ±45 V with an output current of ±15 mA. *OPA*627 gain bandwidth product of 2 MHz is high enough for a closed-loop bandwidth compatible with frequency specifications given by resonance frequencies of devices. Input signal is applied to negative input because of linearity issues and because of a better stability of the circuit; the positive input is grounded. Resistances choice is made according to discrete available values with a tolerance of 0.1%. As all circuit parameters (like loop gain, output impedance, etc.) depend on resistors tolerances, and the lower is resistor tolerance, the higher is circuit performance, a minimum level of 0.1% is wanted. $R_3 = 1\ \mathrm{k}\Omega$ allows to get the maximum desired output current with 10 V at the input, a voltage level common provided by any function generator. The output current thus results:

$$I_{\mathrm{out},L} = -\frac{V_{\mathrm{in_{neg}}}}{R_3} = -\frac{V_{\mathrm{in_{neg}}}}{1\ \mathrm{k}\Omega} \tag{8.10}$$

Resistors $R_1 = R_4 = R_5 = 221\ \mathrm{k}\Omega$ and $R_2 = 220\ \mathrm{k}\Omega$ in order to satisfy condition given by Eq. (8.5). To improve stability of the circuit a $C_C = 2.2$ pF is placed in parallel to R_5 achieving a bandwidth slightly higher than 30 kHz.

8.1.2 A Solution for Differential Driving

As mentioned before, one of the main constraints for driving circuit is to avoid to add unwanted electrostatic forces giving a misleading output signal. The proposed solution to implement a fully differential driving stage is based on the solution shown in [1] and the circuit is revised to specifically improve the feedback network used to keep the MEMS shuttle voltage close to ground.

The generated current is pumped into a MEMS device from one of the two springs ends and an ancillary circuit is used to sample the voltage at this end of springs using a high-impedance buffer and a 180° phase shifted voltage is applied at the other end of the springs through an inverting buffer. Adjustable gain and offset (using two potentiometers) in this loop allow to carefully set the amplitude of anti-phase signal and so to make the MEMS as more balance as possible with respect to ground potential. Figure 8.3 shows final schematic. Non-inverting and inverting buffers use the same operational amplifier as current pump to provide the same range

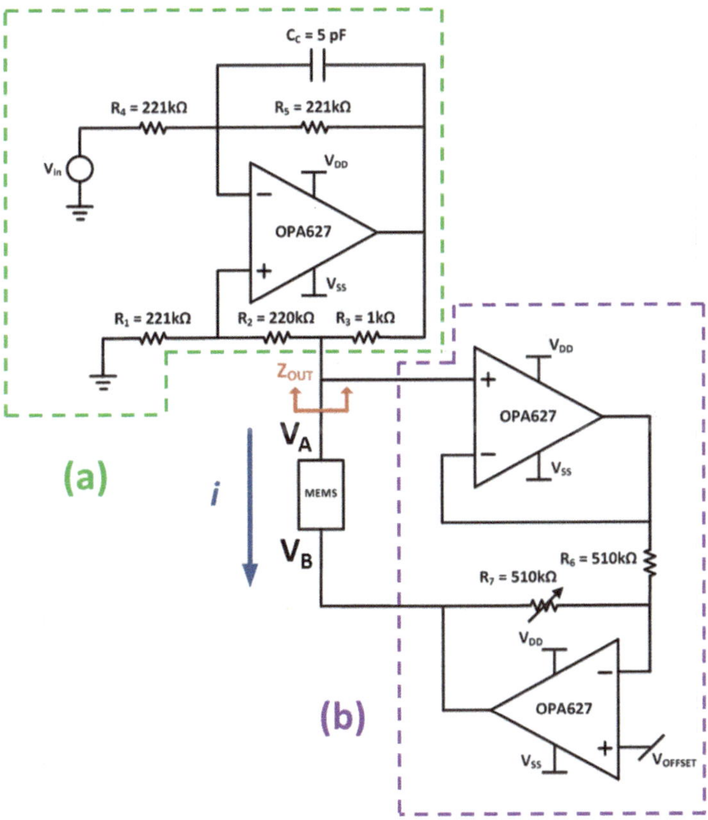

Fig. 8.3 (**a**) Schematic of an improved Howland current pump. (**b**) Additional loop to stabilize the MEMS symmetric with respect to ground potential

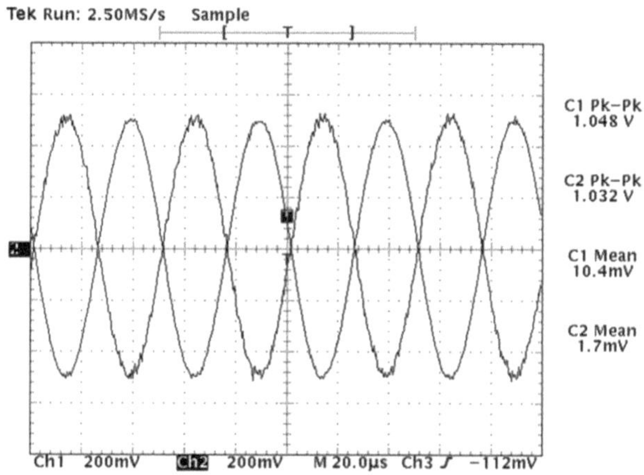

Fig. 8.4 Sinusoidal voltages at MEMS springs ends pumping 500 μA current at 20 kHz

of power supply voltage and capability of driving current. A set of simulations are run using PCB design tool both in time and frequency domain.

8.1.3 PCB Realization and Tests

The circuit is fabricated using a two-layer printed circuit board (PCB). First the driving circuit is tested using a dummy resistor with a known value in order to calibrate the conversion gain (input voltage vs. output current) of the driving stage. The circuit is then applied to a MEMS structure and voltage at two spring ends is monitored. As shown in Fig. 8.4 experimental characterization confirms a good design according to set specifications and the device is almost perfectly balanced driven.

8.2 Capacitance Variations Readout

This section discusses preliminary features about the architecture for readout electronics. In order to validate working principle and to test Lorentz force magnetometers, a discrete components prototype is developed. As already mentioned, a magnetometer can be modeled as two capacitors in series which vary because of a displacement of the moving mass in presence of an external force. Thus, positive or negative capacitance variations can be obtained depending on moving electrode going close or away from the fixed one. As far as readout and driving electronics coupling is concerned, Lorentz force magnetometers case is particularly delicate. With respect to some other typologies of sensors, like accelerometer or gyroscope, magnetometer operating principle requires a current flowing into springs. Accelerometers do not need being driven; possible signals, usually at high frequency to improve SNR, are applied to moving mass exclusively for adopted capacitive readout technique [7]. The more complex case of gyroscopes requires an excitation circuit to keep moving mass oscillating; indeed, their sensitivity depends on moving mass oscillation velocity [8, 9]. In these two cases it is required to couple a voltage signal either to a pad electrically connected to a moving mass (accelerometer) or to an actuator (gyroscope). Magnetometer case has the peculiarity, as said, to need a current flowing in moving mass. This means that the two pads to access moving mass are used to drive a current (as shown in Sect. 5.2, there are overall four pads to connect devices, two of them are used to access moving mass and the two others are for fixed electrodes). Moreover, signal applied to moving mass must avoid to add unwanted electrostatic forces. Summing up, this scenario does not allow to easily use moving mass pads to add other signals. So, a capacitive readout technique which does not need to couple signals to moving mass is preferred to avoid interferences with driving signal. For above-mentioned reasons, solutions proposed in [10–12] are revised with an eye on magnetometer requirements.

In Fig. 8.5 a bonding plan of a device is shown: there are a pad for chip ground (substrate), a pad for stopper biasing (usually kept to GND), two pads for moving mass (*MM*1 and *MM*2) for current generator, and two remaining pads electrically connected to the two stators. In terms of simplicity, only the two stators pads should be used for readout.

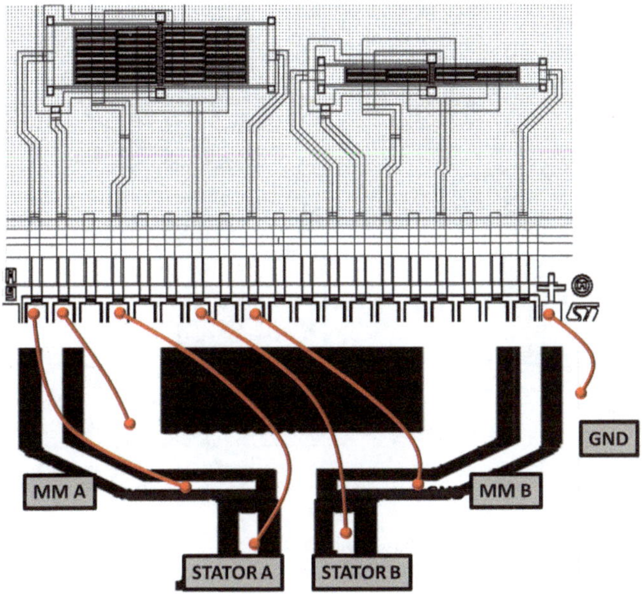

Fig. 8.5 Bonding plan of a $N = 16$ magnetometer. MEMS package is glued on a PCB close to readout amplifiers and directly bonded to PCB paths

Among different solutions which are considered to implement readout electronics is the setup introduced in Sect. 6.2 used for electromechanical characterization. After a set of experimental tests about high frequency test signal and current driving coupling, it turned out that output signal is perturbed a lot.

The choice is limited between a position sense interface based on a voltage readout or on a transimpedance amplifier (TIA) [11]. The first solution is discarded because it suffers from parasitic capacitances affecting the input node, the effect of which can be a complete reduction of signal. A voltage approach is more suitable in case of a single chip solution: MEMS and readout electronics are implemented on the same die and so additional unwanted capacitances can be controlled more accurately. Finally, the chosen solution is based on a transimpedance amplifier taking advantage of the alternated driving signal which provides a capacitance variation right at that working frequency.[1]

A comparison study about signal to noise ratio performance is carried out between a transcapacitance scheme, where current is integrated on a feedback capacitance and a resistor is used for DC path and to discharge the capacitance, and a transresistance scheme where, on the contrary, current is integrated in feedback resistor and a capacitance is used to set closed-loop bandwidth. Schematic for a two channels transimpedance amplifier is shown in Fig. 8.6: values of R_F and C_F vary according to the solution. For a transcapacitance scheme, R_F and C_F must be chosen to set closed-loop pole at a frequency about a decade lower than signal frequency, in case of a transresistance scheme values are chosen to set the pole about a decade higher than driving frequency.

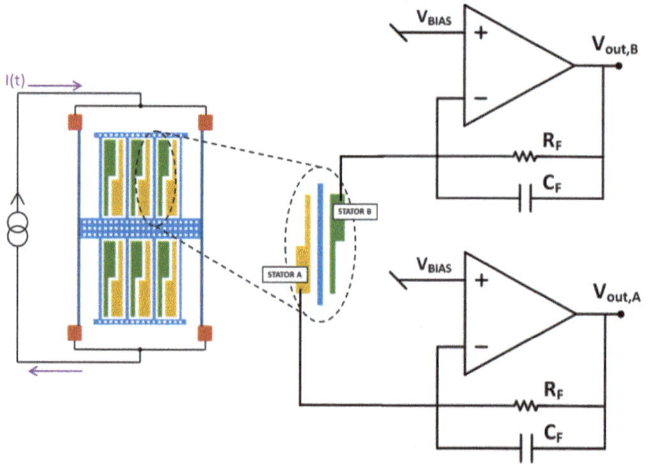

Fig. 8.6 Differential positing sensing based on two single-ended transresistance amplifiers

[1]In order to sense a capacitance variations an AC modulation is required: it can be due either to a capacitance variation or to a voltage variation.

Generally, current flowing across a capacitor with a voltage V is:

$$I(t) = \frac{dQ(t)}{dt} = C(t) \cdot \frac{dV(t)}{dt} + V(t) \cdot \frac{dC(t)}{dt} . \quad (8.11)$$

If V is a constant voltage the first term is null and current only depends on capacitive variations, which are at resonance frequency. In presence of a variable MEMS capacitance, its value can be written as:

$$C = C_0 \left[1 + \frac{\Delta C_0}{C_0} \cdot \sin(2\pi \cdot f_r \cdot t) \right] , \quad (8.12)$$

under the hypothesis that external force is sinusoidal at frequency f_r. Differentiating previous equation the following capacitance variation is obtained:

$$\Delta C = 2\pi \cdot f_r \cdot \Delta C_0 \cdot \cos(2\pi \cdot f_r \cdot t) , \quad (8.13)$$

which can be put in Eq. (8.11) to get a current:

$$I(t) = V_{\text{BIAS}} \cdot \Delta C_0 \cdot \cos(2\pi \cdot f_r \cdot t) \cdot 2\pi \cdot f_r , \quad (8.14)$$

V_{BIAS} being DC voltage across the capacitor.

Output signals for the two schemes, transcapacitance and transresistance respectively, are:

$$\begin{aligned} V_{\text{out}_{\text{tc}}}(t) &= I(t) \cdot \left| \frac{1}{sC_F} \right|_{f=f_r} \\ &= \frac{V_{\text{BIAS}} \cdot \Delta C_0 \cdot 2\pi \cdot f_r}{2\pi \cdot f_r \cdot C_F} \cdot \cos(2\pi \cdot f_r \cdot t) \\ &= V_{\text{BIAS}} \cdot \frac{\Delta C_0}{C_F} \cdot \cos(2\pi \cdot f_r \cdot t) , \end{aligned} \quad (8.15)$$

and

$$\begin{aligned} V_{\text{out}_{\text{tr}}}(t) &= I(t) \cdot R_F \\ &= V_{\text{BIAS}} \cdot \Delta C_0 \cdot 2\pi \cdot f_r \cdot R_F \cdot \cos(2\pi \cdot f_r \cdot t) . \end{aligned} \quad (8.16)$$

The oscillating term is not considered any longer since it only gives an information about the signal frequency, at resonance, but it does not affect signal amplitude, in which the information about magnetic field is. Low-frequency ($\approx$10 Hz) magnetic field information is an AM modulation of this signal at resonance frequency. Indeed, it is supposed to sample signal at peaks with a demodulating scheme. For an immediate comparison between the two schemes SNR, operational amplifiers noise is not considered as it is equal for both cases and supposed negligible; only feedback resistor noise is considered resulting, at the output, in:

$$S_{R_F,n} = 4 \cdot k_{\mathrm{B}} \cdot T \cdot R_F \quad [\mathrm{V}^2/\ \mathrm{Hz}] \,. \tag{8.17}$$

Supposing to filter signals with a narrow bandwidth around the working frequency, SNR of a transcapacitance scheme is:

$$\begin{aligned} \left(\frac{S}{N}\right)^2_{\mathrm{tc}} &= \frac{\left(V_{\mathrm{BIAS}} \cdot \frac{\Delta C_0}{C_F}\right)^2}{\frac{4 \cdot k_{\mathrm{B}} \cdot T}{R_F} \cdot \left|\frac{1}{sC_F}\right|^2_{f=f_r} \cdot \mathrm{BW}} \\ &= \frac{V^2_{\mathrm{BIAS}} \cdot \Delta C_0{}^2 \cdot R_F \cdot (2\pi \cdot f_r)^2}{4 \cdot k_{\mathrm{B}} \cdot T \cdot \mathrm{BW}} \,, \end{aligned} \tag{8.18}$$

$$\begin{aligned} \left(\frac{S}{N}\right)^2_{\mathrm{tr}} &= \frac{(V_{\mathrm{BIAS}} \cdot \Delta C_0 \cdot 2\pi \cdot f_r \cdot R_F)^2}{4 \cdot k_{\mathrm{B}} \cdot T \cdot R_F \cdot \mathrm{BW}} \\ &= \frac{V^2_{\mathrm{BIAS}} \cdot \Delta C_0{}^2 \cdot R_F \cdot (2\pi \cdot f_r)^2}{4 \cdot k_{\mathrm{B}} \cdot T \cdot \mathrm{BW}} \,. \end{aligned} \tag{8.19}$$

Assuming to filter the signal around the carrier using a narrow band filter (about few tens of Hz according to the specifications) the two schemes provide the same signal to noise ratio. The chosen solution is based on a transresistance amplifier. Discrete components electronics are used for a preliminary characterization of devices and magnetic field measure, then a VLSI circuit is implemented for the final demonstrative prototype.

8.3 MEMS and Electronics Coupling: Parasitics Reduction

As detailed in Sect. 8.5.1, magnetic field measures are made using a specific instrument for magnetic field generation. In a first approach discrete electronics were on a different PCB with respect to the sensor being more versatile to test several devices. The two boards are connected using shielded cables.

This solution shows two main drawbacks:

- The use of shielded cables adds additional parasitic capacitances at the input node, the virtual ground of transresistance amplifier. Main effects are a reduction of stability and bandwidth. In addition, measures were not reliable because of the mechanical instability of the two boards and signals coupling to virtual ground through cables.
- Coupling of signals among PCB traces coming from the position where MEMS is bonded: two traces are for driving signal and two for readout (see Fig. 8.7a).

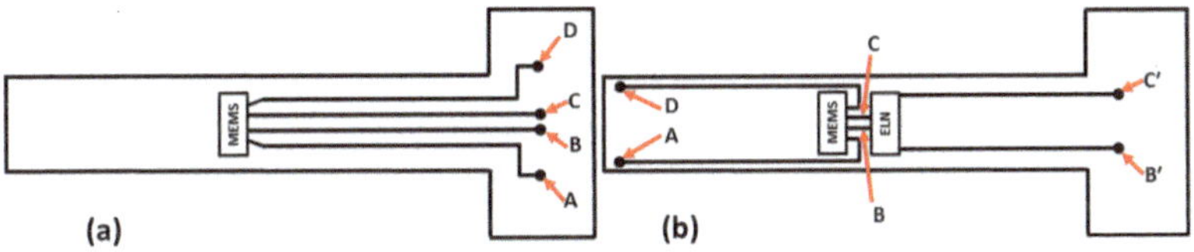

Fig. 8.7 PCB hosting MEMS and discrete components electronics. (**a**) First type with driving paths close to sensing paths. (**b**) Improved layout to minimize couplings. PCB shape is constrained by magnetic field generator socket

Fig. 8.8 Details of PCB where MEMS is glued and bonded to readout electronics. Two short traces connect the bonding pads to input of transresistance amplifiers and two other copper paths are traced in the opposite direction to a connector for driving stage

Figure 8.7b reports an adopted solution to minimize couplings among PCB paths. Paths A and D for current driving are routed to the opposite side of the PCB increasing the relative distance between each others. Readout electronics (a detailed description of the design is reported in Sect. 8.4) is put close to the MEMS in order to reduce parasitic capacitances and disturbs. Signals are provided at the output of this PCB after a first amplification and so improving signals immunity.

In Fig. 8.8 a close-up photograph of the implemented solution is shown: MEMS is bonded very close to the readout amplifier to reduce parasitic capacitances and improve signal to noise ratio.

8.4 Transresistance Amplifier

The implemented solution for discrete components readout electronics consists of two channels transresistance amplifier: current signal with its information about capacitance variation is transduced into a voltage signal through the feedback resistor. This voltage signal can thus be further amplified, demodulated, and sampled. For earth magnetic field measurements detected signal is usually low, corresponding to a current on the order of pA, therefore a low-noise frontend is fundamental. Transresistance amplifier is implemented using an operational amplifier in a closed-loop with a resistor; thanks to the negative feedback this circuit is used both to sense the current and to bias MEMS stator with voltage applied to positive input.

The design of the transresistance amplifier is based on two preliminary constraints: (1) feedback resistor value must be as large as possible to reduce input current noise but at the same time signal bandwidth must be guaranteed; a few MΩ resistor could be in the right range not to excess an input current noise of $\text{pA}/\sqrt{\text{Hz}}$. (2) operational amplifier package must be tiny, possibly a 5 pins type to be able to put two stages close to the MEMS. So a suitable operational amplifier is chosen among commercially available products taking into account the abovementioned requirements and the fact that the noise floor of the closed-loop amplifier should not be affected by opamp contributions. Then the precise value of feedback resistor is chosen to satisfy bandwidth and stability requirements.

Operational amplifier *AD*8065 by Analog Devices [13] is chosen mainly because of its low input parasitic capacitances, low noise, and tiny package, $SOT-23$ 5-pins. It is a high-performance 145 MHz FET inputs opamp with input referred noise:

$$\begin{aligned} \sqrt{S_{V,n}} &= 7\ \text{nV}/\sqrt{\text{Hz}}\ @10\ \text{kHz} \\ \sqrt{S_{I,n}} &= 0.6\ \text{fA}/\sqrt{\text{Hz}}\ @10\ \text{kHz}\,. \end{aligned} \tag{8.20}$$

Average leakage current is about 6 pA and input offset is 1.5 mV; common mode rejection ratio is CMRR $=$ 100 dB. In order to satisfy bandwidth requirement resistor and capacitance chosen values are:

$$\begin{aligned} R_F &= 3\ \text{M}\Omega \\ C_F &= 200\ \text{fF}\,, \end{aligned} \tag{8.21}$$

resulting in a signal bandwidth of:

$$\text{BW} = \frac{1}{2\pi R_F C_F} = 265\ \text{kHz}\,. \tag{8.22}$$

Voltage at positive input of operational amplifier is set using an SMD potentiometer whose value of 500 Ω is chosen not to degrade noise performance of the stage.

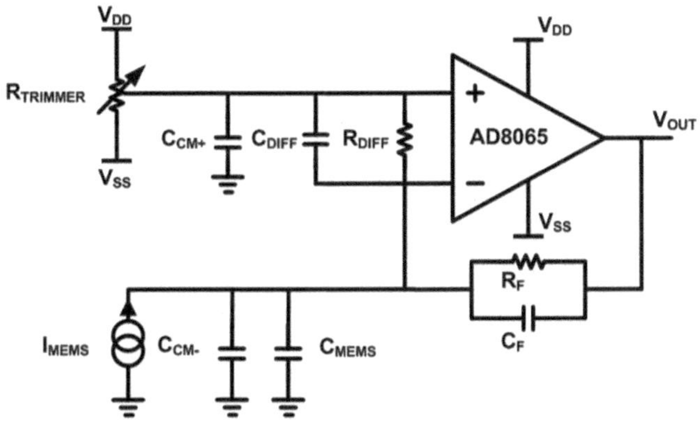

Fig. 8.9 Transresistance schematic with parasitic capacitances and resistances due to operational amplifier

8.4.1 Frequency Response

The ideal transfer function of the amplifier is:

$$T_{\mathrm{id}}(s) = \frac{V_{\mathrm{out}}(s)}{I_{\mathrm{in}}(s)} = \frac{R_F}{1 + s \cdot R_F \cdot C_F}. \tag{8.23}$$

In order to make a stability analysis of the amplifier, loop gain and the consequent phase margin must be studied considering singularities both of the opamp open-loop gain and given by feedback parts. In Fig. 8.9 a schematic of the amplifier is shown with input resistors and parasitic capacitances which have an important role on the stability. In first approximation R_{trimmer} can be neglected because it is a relative low value compared to differential resistance of the opamp. Under this hypotheses also common mode resistance and capacitance, $R_{\mathrm{CM,+}}$ and $C_{\mathrm{CM,+}}$, can be neglected as they are shorted to ground. Differential resistance and capacitance, R_{DIFF} and C_{DIFF}, still play a role in loop gain. Even though the introduced assumptions allow a simpler analytical evaluation of the stability, a circuit simulator is used for a complete stability analyses including higher frequency singularities. In the analysis of loop gain C_{MEMS} is considered equal to maximum estimated value with an extra for additional parasitics due to bondings and chip internal paths. $R_{\mathrm{DIFF}} \sim$ GΩ is so large that it can be neglected. Thus, loop gain can be written as follows:

$$G_{\mathrm{loop}} = -A(s) \cdot \frac{1 + s\tau_z}{1 + s\tau_p}, \tag{8.24}$$

where $A(s)$ is *AD*8065 open-loop gain (Fig. 8.10), τ_z and τ_p are time constants of zero and pole, respectively. Their corresponding frequencies are:

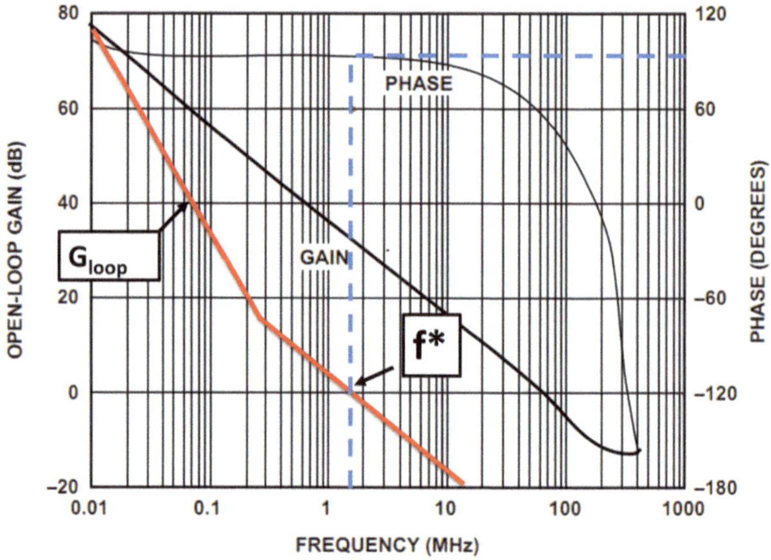

Fig. 8.10 Open-loop gain of *AD*8065 used to estimate TIA loop gain

$$f_z = \frac{1}{2\pi \cdot \tau_z} = \frac{1}{2\pi \cdot R_F \cdot C_F} = 265 \text{ kHz} \,;$$

$$f_p = \frac{1}{2\pi \cdot \tau_p} \tag{8.25}$$

$$= \frac{1}{2\pi \cdot R_F \cdot (C_F + C_{\text{MEMS}} + C_{\text{CM}-} + C_{\text{diff}})} = 10 \text{ kHz} \,,$$

considering a conservative value of about 5 pF for MEMS capacitance and parasitic arising from bondings, sockets, and connections.

Knowing the position of loop gain singularities, an estimation of phase margin can be obtained:

$$\varphi_m = 180^\circ - \varphi(A(s))|_{f=f^*} - \arctan\left(\frac{f^*}{f_p}\right) + \arctan\left(\frac{f^*}{f_z}\right) \approx 78^\circ \,, \tag{8.26}$$

$f^* = 1.6$ MHz being the frequency where $G_{\text{loop}}(s)$ intersects 0 dB axes. The designed amplifier has quite a good stability.

8.4.2 Noise Analysis

This section deals with a detailed analyses of noise contributions in the transresistance amplifier considering the following sources:

- Thermal noise of the resistor R_F;
- Current noise at the positive and negative inputs of the operational amplifier;
- Op amp voltage noise;
- Thermal noise of the biasing potentiometer.

They can be transferred to TIA output as follows:

$$R_F : S_{\text{out},R_F} = \frac{4k_{\text{B}}T}{R_F} \cdot \left(\frac{R_F}{1 + s \cdot R_F C_F}\right)^2 \tag{8.27}$$

$$I_{n+} : S_{\text{out},I_{n+}} = S_{\text{in},I_{n+}} \cdot R^2_{\text{trimmer}} \left(\frac{1 + s \cdot R_F \cdot (C_F + C_{\text{MEMS}} + C_{\text{Par}})}{1 + s \cdot R_F C_F}\right)^2 \tag{8.28}$$

$$I_{n-} : S_{\text{out},I_{n-}} = S_{\text{in},I_{n-}} \cdot \left(\frac{R_F}{1 + s \cdot R_F C_F}\right)^2 \tag{8.29}$$

$$V_n : S_{\text{out},V_n} = S_{\text{in},V_n} \cdot \left(\frac{1 + sR_F \cdot (C_F + C_{\text{MEMS}} + C_{\text{Par}})}{1 + s \cdot R_F C_F}\right)^2 \tag{8.30}$$

$$R_{\text{trimmer}} : S_{\text{out},R_{\text{trimmer}}} = 4k_{\text{B}}T \cdot R_{\text{trimmer}} \cdot \left(\frac{1 + sR_F \cdot (C_F + C_{\text{MEMS}} + C_{\text{Par}})}{1 + s \cdot R_F C_F}\right)^2 , \tag{8.31}$$

where C_{Par} accounts for op amp parasitics.

Figure 8.11 reports noise densities at the output of TIA using a two pole approximation in their transfer functions. R_F noise is dominant in the range of frequencies of MEMS devices and its value is:

$$\sqrt{S_{R_F,n}} = \sqrt{(4 \cdot k_{\text{B}}T \cdot R_F)} = 223 \text{ nV}/\sqrt{\text{Hz}} . \tag{8.32}$$

The theoretical study is supported by circuit simulations and proved by experimental measures performed using Spectrum Analyzer *HP*4195*A*. Figure 8.12 shows the measured output voltage noise, instrument noise floor, and feedback resistor theoretical noise.

In order to get an estimation of the minimum detectable signal using the implemented stage, output noise value can be converted into an equivalent input capacitance noise variation; noise contribution of the two channels (uncorrelated) must be taken into account and the estimated resolution is:

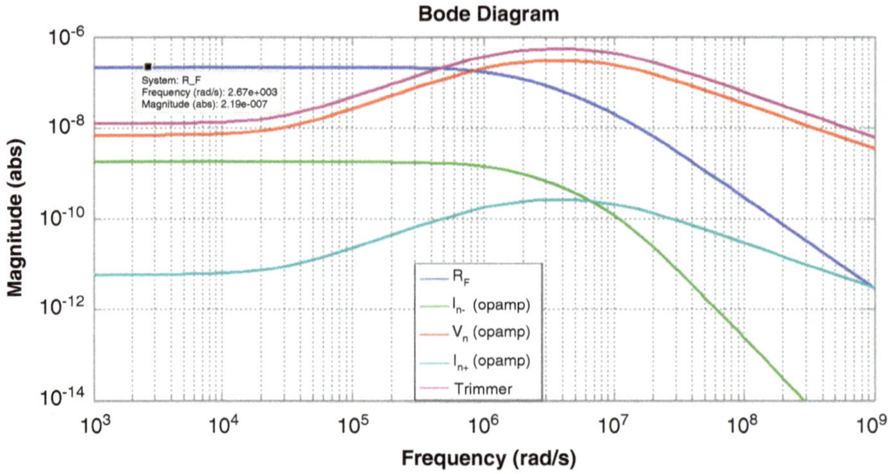

Fig. 8.11 Simulated spectral noise densities at transimpedance output

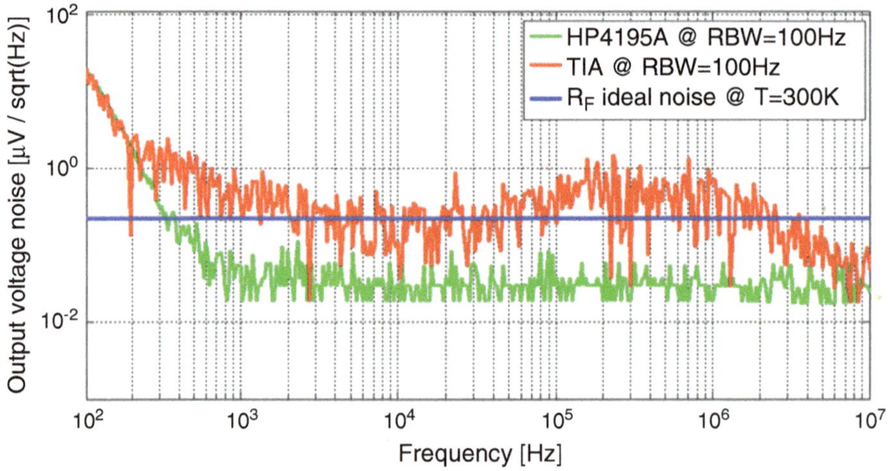

Fig. 8.12 Measured output noise (*red line*); instrument noise floor (*green line*) and theoretical resistor thermal noise associated to R_F @ 300 K (*blue line*)

$$\sqrt{S_{C_F,n}} = \frac{\sqrt{2 \cdot S_{R_F}}}{2\pi \cdot f_r \cdot R_F \cdot V_{\mathrm{BIAS}}} = 0.2\ \mathrm{aF}/\sqrt{\mathrm{Hz}} \tag{8.33}$$

From this, noise spectral density in terms of magnetic field can be estimated to be 0.97 $\mu\mathrm{T}/\sqrt{\mathrm{Hz}}$. This value is obtained under the hypothesis of using a $N = 4$ magnetometer in the following working condition:

- $V_{\text{BIAS}} = 3$ V;
- $I_{\text{Lorentz}} = 200$ μA$_{\text{rms}}$;
- $f_r = 28.3$ kHz (theoretical value).

Minimum magnetic field detectable value depends on the bandwidth used to integrate the noise spectral density; using a lock-in based demodulation technique a narrow filtering band (e.g., BW = 15 Hz) can be used around the working frequency.

8.4.3 *Differential to Single-Ended Conversion*

Even though signal to noise ratio is set by first electronic stage, a further amplification is convenient before feeding this signal to an external lock-in amplifier. In addition to amplification, a conversion at hardware level of these two channels into one single-ended signal containing information about differential capacitance variation is more comfortable for further processing. For this purpose an Instrumentation Amplifier (INA) is chosen because of its properties to amplify the difference between the two input signals and to reject common mode. Many different high-performance INAs are commercially available: Texas Instrument INA129 is chosen thanks to its noise, bandwidth, and CMRR features:

$$\begin{aligned} \text{CMRR} &= 106\,\text{dB} \quad @G = 10 \\ \sqrt{S_{V,\text{INA}}} &= 8\ \text{nV}/\sqrt{\text{Hz}} \quad @1\ \text{kHz} \\ \sqrt{S_{I,\text{INA}}} &= 0.3\ \text{pA}/\sqrt{\text{Hz}} \quad @1\ \text{kHz}\,. \end{aligned} \tag{8.34}$$

Using an external resistor, $R_G = 1$ kΩ gain is set to:

$$G_{\text{INA}} = 1 + \frac{49.4\ \text{k}\Omega}{R_G} = 51\,. \tag{8.35}$$

Reference pin of INA129 is used to high-pass filter using an integrator. Figure 8.13 shows a simplified schematic.

8.5 Magnetic Field Measures

8.5.1 *Experimental Setup*

Magnetic field measurements are performed with a magnetic field compensator and generator *Palm Gauss PG-5G magnetic field canceller* by *Aichi*. This instrument, based on a three-axial system of Helmholtz coils and equipped with reference

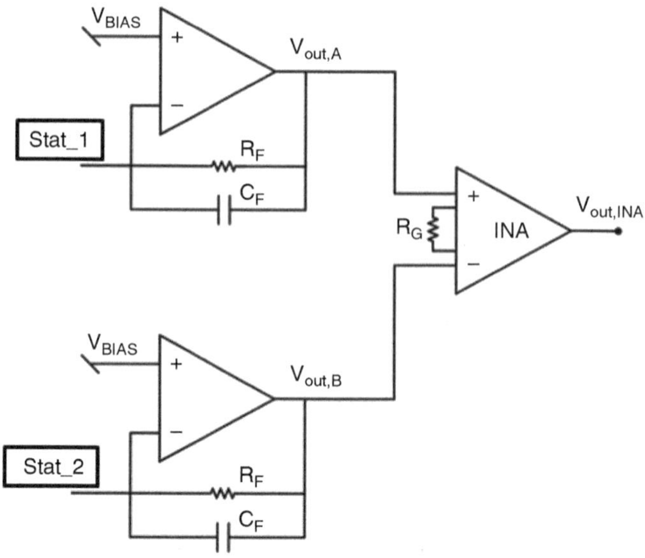

Fig. 8.13 Simplified schematic of the instrumentation amplifier used to amplify and convert two channels into a single-ended signal

magnetometers, is capable of precisely measuring the magnetic field in the volume enclosed by the coils, and compensating it. The main characteristics of this instrument are:

- Magnetic field control direction: three axis (*X*-*Y*-*Z*);
- Effective control space: ±70 mm × ±70 mm × ±70 mm;
- Maximum magnetic field cancellation: ±450 μT;
- Magnetic field inside cancelled area: <1 μT;
- Rated resolution: 100 nT;
- Helmholtz coil dimensions:
  X-axis 481 × 481 × 265 mm
  Y axis 542 × 542 × 299 mm
  Z axis 420 × 420 × 230 mm

For measurements automation, data are collected using Data Acquisition Board (DAQ) NI PCI-6281 by National Instruments™ and a control and data acquisition software is implemented using LabVIEW. The main features of this board are 18 bit resolution and 625 kS/s sampling rate for each analog input channel. LabVIEW software is used for an initial calibration of reference sensors and to implement a control loop to set a magnetic field. Indeed $PG - 5G$ magnetic field generator does not have an intrinsic closed-loop integrated in its control unit and driving is done by setting the value of current flowing in each coil. It is thus important to implement a software loop to set a current in one or more directions, to sense the corresponding imposed magnetic field, to compare it with a reference value set by the user and

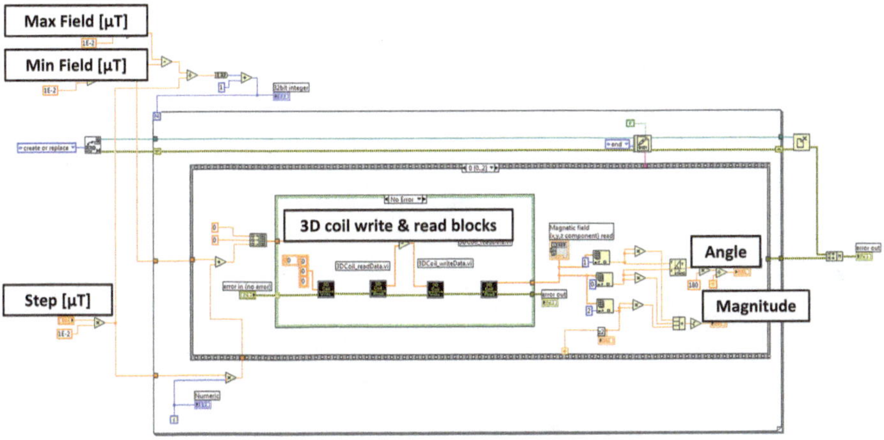

Fig. 8.14 A sub-block of LabVIEW software to drive $PG-5G$ magnetic field canceller. User can set a minimum value, a maximum value, and a step for the magnetic field sweep

consequently acting on current flowing in coils. After a few iterations the desired values are usually obtained in magnetic field controlled area (Fig. 8.14).

A preliminary calibration and test of the available setup is done measuring the acquisition noise and the stability of generated magnetic field. The measured noise due to acquisition system is $\approx$105 μV_{rms} which is much higher than the only contribution due to theoretical quantization noise, given by:

$$\sigma^2_{DAQ} = \frac{\mathrm{LSB}^2}{12} = \frac{\mathrm{FSR}^2}{2^n \cdot 12} = 2.2\ \mu\mathrm{V} \tag{8.36}$$

being FSR = 2 V and $n = 18$.

After $PG - 5G$ instrument calibration, a region of nearly uniform magnetic field of a known value can be generated with an offset value of 0.33 μT and a standard deviation of 80 nT which is much lower than magnetic field step used during measures, 0.5 μT or larger.

8.5.2 *Sensitivity*

The MEMS magnetometer under test, carefully glued and wire bonded to the boards with the driving and readout electronics, is placed at the center of the instrument to avoid side effects due to field non-uniformity near the coils (see Fig. 8.15). For each measurement the magnetometer is driven at its resonance frequency with a peak current of 125 μA per spring (i.e., 250 μA of overall peak current is delivered by the Howland pump). In always-on operation mode, this corresponds to a power dissipated in the springs $P_{\mathrm{Joule}} = 83\ \mu\mathrm{W}$, considering a typical MEMS resistance in

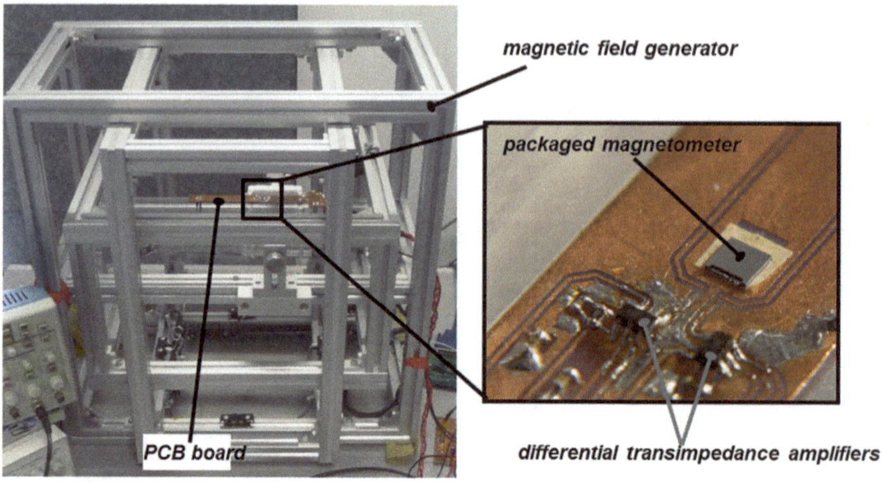

Fig. 8.15 The Aichi magnetic field canceller with the PCB hosting the magnetometer. The system of tri-axial coils for the calibration of the magnetic field is clearly visible. The close-up shows how the magnetometer was glued directly on the PCB board very close to the differential transimpedance stage and placed inside the uniform magnetic field region

the range $R_{\text{MEMS}} = 2.5–3\ \text{k}\Omega$. The instrument first compensates the environmental magnetic field and then generates a stair of increasing field values for the Z-axis from -100 to 100 μT, by steps of 0.5 μT (keeping nominally null the magnetic field over the other directions).

As already introduced, the instrument is controlled through a *LabVIEW* program that also samples, through a data acquisition module (National Instruments™ NI PCI-6281), the output value at each point in the ramp. An example of the measurement for the smallest designed device ($N = 4$) is reported in Fig. 8.16: the voltage output of the acquisition setup is reported on the y axis as a function of the field value on the x-axis, together with the best linear fitting after an offset compensation. Offset is typically in the order of ≈4 mV, corresponding to ≈27 μT, in good agreement with the magnetic field offset measured during calibration (and so using reference sensors). The obtained overall sensitivity is around 150 μV/μT, in line with the theoretical predictions of 195 μV/μT for a gap $g = 2.1$ μm. Differences might result from a process under-etch with respect to the expected one. The results obtained for the other devices—having a different number of parallel-plate cells—are similar to the one shown here, in agreement with the modelling theory and the results obtained for the quality factors reported in Sect. 7.1.2. Performance differences in the order of ±5% are observed on devices belonging to different batches; these variations are in line with measured differences in the resonance frequency, likely due to the process variance obtained in productions specifically done for research puroposes.

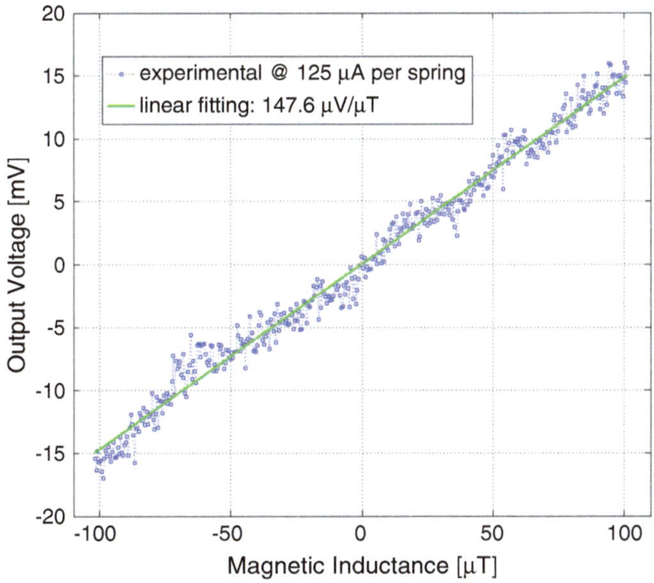

Fig. 8.16 Output voltage as a function of the magnetic field in the Z-direction (*blue square markers*) with its best linear fitting (*green line*)

8.5.3 *Resolution*

The resolution was estimated as the standard deviation of a set of 100,000 sampled measurements at the output of the system for an arbitrary fixed value of the magnetic field. Measurements were repeated using different low-pass filtering bandwidths of the lock-in amplifier in the range 1 Hz–2 kHz. The obtained result is shown in Fig. 8.17: a comparison with the resistors noise predictions shows a fairly good agreement for the highest bandwidth. At low-frequency values, some more noise can be observed whose origin may be either in the driving oscillator noise or in the $1/f$ noise of the lock-in amplifier itself. The obtained results, converted in terms of magnetic field through the measured sensitivity, give a minimum measurable magnetic field of 520 $\mathrm{nT \cdot mA/\sqrt{Hz}}$; this means that a sub-$\mu\mathrm{T}/\sqrt{\mathrm{Hz}}$ resolution can be obtained with less than 1 mA of overall driving current.

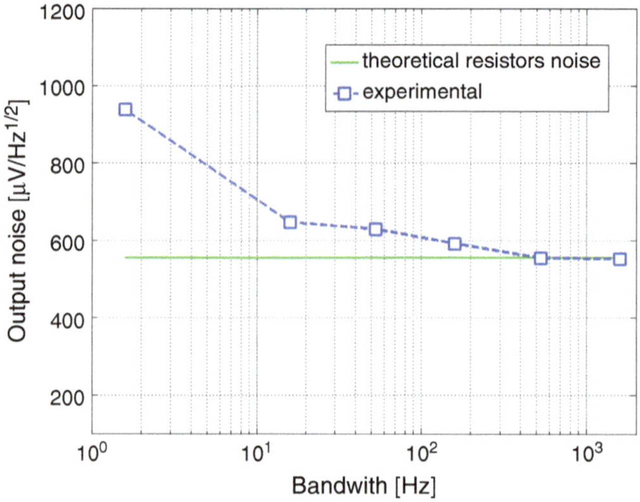

Fig. 8.17 Measured resolution as a function of the measurement bandwidth (*blue square markers*) and comparison with the theoretical noise of the resistors

8.5.4 Influence of Possible Fluctuations of the Damping Coefficient

Two main sources of possible variations in the damping coefficient have been considered, to verify their influence on the device mechanical sensitivity, resolution, and bandwidth.

- *Process changes from part to part*: for the structures with $N = 4$, the measured quality factor at room temperature (about 22 °C) was observed to vary between 335 and 350 for six measured samples. Only a seventh device showed a significantly different Q (around 372); this might be due to a corner device where overetch is not in the typical range.
- *changes of the damping coefficient with temperature*: this phenomenon, related to the changes of the pressure with temperature within a constant volume package, is well described, for instance, in [14]. In this reference the authors suggest a dependence of the quality factor on the temperature with a $1/T^{\gamma}$ law, γ being close to 0.5 in case of air damping. Measurements of the quality factor with respect to temperature [15, 16] were performed using the *M/N/TY55* climatic chamber from *Angelantoni Climatic Systems*, to verify the temperature behavior in the presented devices. The results are shown in Fig. 8.18. In these experiments the quality factor shows a variation from 340 to 312 for a measurement range between 5° and 45°. It can be thus evidenced a percentage variation of the

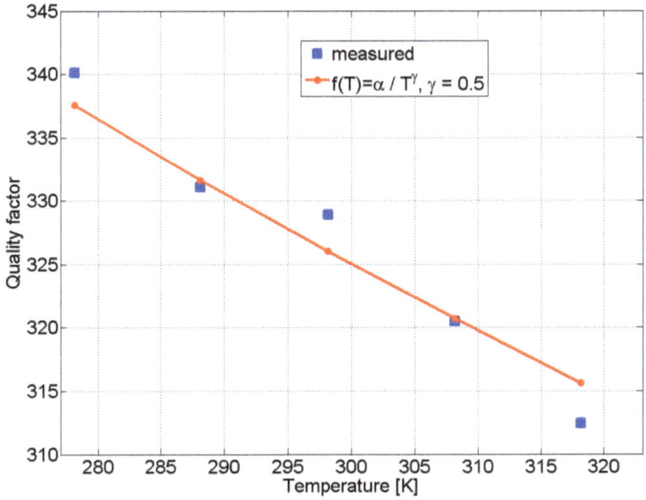

Fig. 8.18 Experimental measurement of the quality factor with respect to the absolute temperature for the device with $N = 4$

quality factor around 0.2%/K. Within this range, the measured device resonance frequency varies less than 20 ppm/K, confirming that the changes in the Q are mostly related to changes in the pressure.

Considering the typical operation range of consumer devices (−40° to 85°), it turns out that, among the two possible sources of fluctuation of the damping coefficient, the most relevant one is its dependence on the temperature (25% change on the whole range). When discussing large volume applications, one should however consider that all the parts are generally subject to an initial calibration: this compensates for the native changes from part to part due to the process variance. One should moreover take into account that a temperature sensor is often embedded in the same ASIC which includes the MEMS readout electronics: this can be exploited in operation for post-acquisition compensation of temperature changes. Therefore, initial calibration and temperature compensations similar to those applied, for instance, to a gyroscope (such a device is based on resonant driving and sensing too) will be mandatorily required also for the magnetometers. Note that these post-production operations correct digitally for the overall device gain: issues that still remain unchanged, even after the suggested compensations, are the change in the intrinsic resolution (due to the change in the native mechanical sensitivity) and the change in the open-loop bandwidth (due to the change in the quality factor). The former degrades at high temperatures, the latter degrades at low temperatures.

References

1. J. Kyynäräinen, J. Saarilahti, H. Kattelus, A. Kärkkäinen, T. Meinander, A. Oja, P. Pekko, H. Seppä, M. Suhonen, H. Kuisma, S. Ruotsalainen, M. Tilli, Sens. Actuators, A **142**(2), 561 (2008). doi:10.1016/j.sna.2007.08.025
2. R.A. Pease, A comprehensive study of the Howland current pump. Application Note 1515. National Semiconductor (2008)
3. R. Stiz, P. Bertemes, A. Ramos, V. Vincence, Latin America Transactions. IEEE Lat. Am. Trans. **7**(5), 514 (2009). doi:10.1109/TLA.2009.5361187
4. P. Pouliquen, J. Vogelstein, R. Etienne-Cummings, in *IEEE Biomedical Circuits and Systems Conference, 2008. BioCAS 2008* (2008), pp. 33–36. doi:10.1109/BIOCAS.2008.4696867
5. J. Steele, T. Green, *Proper Compensation Techniques Will Let The Improved-Howland And Floating-Load Current Sources Perform Up To Par. Tame Those Versatile Current-Source Circuits* (Apex Microtechnology Corp., Tucson, 1992)
6. G. Hammond, C. Speake, M. Stiff, Int. J. Electron. **95**(4), 351 (2008). doi:10.1080/00207210801976503
7. M. Lemkin, B. Boser, IEEE J. Solid State Circuits **34**(4), 456 (1999). doi:10.1109/4.753678
8. J. Geen, S. Sherman, J. Chang, S. Lewis, IEEE J. Solid-State Circuits **37**(12), 1860 (2002). doi:10.1109/JSSC.2002.804345
9. A. Sharma, M. Zaman, F. Ayazi, IEEE J. Solid-State Circuits **42**(8), 1790 (2007). doi:10.1109/JSSC.2007.900282
10. B. Boser, in *1997 International Conference on Solid State Sensors and Actuators, 1997. TRANSDUCERS '97 Chicago*, vol. 2 (1997), pp. 1169–1172. doi:10.1109/SENSOR.1997.635413
11. J. Wu, G. Fedder, L. Carley, IEEE J. Solid-State Circuits **39**(5), 722 (2004). doi:10.1109/JSSC.2004.826329
12. M. Saukoski, L. Aaltonen, K. Halonen, T. Salo, in *IEEE International Symposium on Circuits and Systems, 2005. ISCAS 2005*, vol. 6 (2005), pp. 5377–5380. doi:10.1109/ISCAS.2005.1465851
13. A. Devices, Datasheet (2010)
14. B. Kim, M. Hopcroft, R. Candler, C. Jha, M. Agarwal, R. Melamud, S. Chandorkar, G. Yama, T. Kenny, J. Microelectromech. Syst. **17**(3), 755 (2008). doi:10.1109/JMEMS.2008.924253
15. R. Dean, S. Castro, G. Flowers, G. Roth, A. Ahmed, A. Hodel, B. Grantham, D. Bittle, J. Brunsch, IEEE Trans. Ind. Electron. **58**(7), 2591 (2011). doi:10.1109/TIE.2010.2070772
16. C. Buffa, A. Tocchio, G. Langfelder, IEEE Trans. Instrum. Meas. **61**(7), 2012 (2012). doi:10.1109/TIM.2012.2182866

Chapter 9
ASIC Design

This chapter deals with the final steps for a prototype development, according to the methodology introduced in Sect. 2.1:

12 Study of a suitable architecture to be integrated into an ASIC, based on results obtained from validation of concept with discrete components electronics.
13 Design and production of a low-noise position sense interface based on a standard CMOS technology.
14 Experimental characterization of a multi-chip prototype.

9.1 Architecture

In order to develop a final system for detection of earth magnetic field some more considerations must be taken into account to design an ASIC with respect to the first demonstrative prototype based on discrete components, as shown in Chap. 8. First set of measures followed an accurate phase of electromechanical characterization of devices useful to make a selection of fully working parts and detect their resonance frequencies. MEMS are then glued close to readout electronics and used for magnetic field measures. Thus, it is known for each device the precise frequency to be used for current excitation at resonance. This procedure works well during a research and prototyping phase but it is affordable to follow the same procedure for industrialization and mass production, due to amount of parts to be characterized and related costs.

These remarks lead to the need to evaluate whether a closed-loop or an open-loop control is preferable for the final two chips system [1]. The first control type, closed-loop, requires a feedback network to drive the driving block and adjust the driving frequency around device resonance frequency. In this way, neglecting variation which can be due to device to device quality factor variation, any device is driven at the correct frequency always providing the maximum output signal and so

C. Buffa, *MEMS Lorentz Force Magnetometers*,
DOI 10.1007/978-3-319-59412-5_9

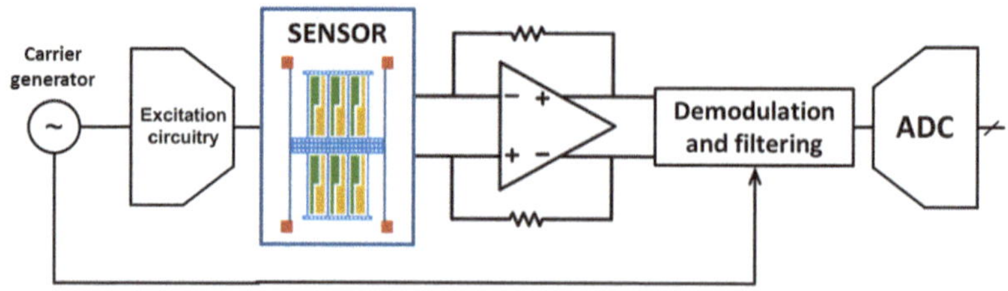

Fig. 9.1 Simplified block diagram of an open-loop implementation. Oscillator signal is used both to drive excitation circuit and as a reference for a synchronous demodulation filtering technique

guaranteeing a relatively good repeatability among samples. On the other hand, in an open-loop system, driving frequency must be set directly tuning oscillator parts or, in case a digital to analog converter (DAC) is used for excitation, storing the correct value of driving frequency in a memory (i.e., with an initial calibration after production; any sensor usually has an initial calibration/self-test after production).

The choice about loop typology is mainly made according to power consumption considerations, as they represent an extremely important specification for systems to be embedded in battery powered devices. Main blocks of driving and readout electronics (at least as far as frontend is concerned) are in common for both open-loop and closed-loop control. This latter has a higher power consumption due to the feedback network to tune oscillation frequency, in addition to a larger ASIC occupied area. Taking into account this general system level remarks, an open-loop architecture is implemented (schematically shown in Fig. 9.1). The problem related to driving frequency tuning still remains.

As emerged with electromechanical tests, even once the magnetometer type is chosen, it is not reliable to set a priori a driving frequency. To this problem, two possible solutions can be proposed: an auxiliary circuitry measures the device resonance frequency either after production as final test before commercialization or when the compass application is executed. Even in this second case, the additional power consumption related to self-test would not affect the overall power consumption during application execution. In this way, the main problem related to open-loop control could be overcome.

After deciding control system type, frontend electronics must be chosen. As sensors are based on a differential capacitive scheme, a differential readout electronic architecture is consequently chosen. Analysis reported in Sect. 8.2 still applies to ASIC development with the exception that a fully differential operational amplifier should be used (instead of two single-ended stages) to minimize phase shift between the two channels, to reduce noise contribution, and to avoid active gain stages mismatches which might affect the overall closed-loop transfer function of positive and negative inputs.

In case of Lorentz force magnetometers driving current signal, which is at resonance, represents a "high-frequency" carrier which is amplitude modulated by capacitance variation and so by magnetic field. With an approach based

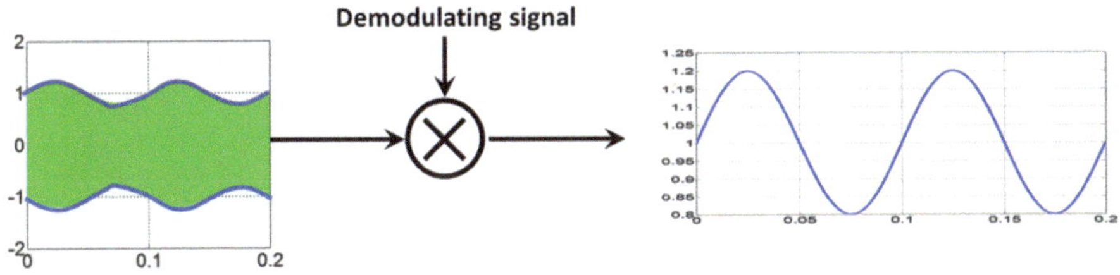

Fig. 9.2 Schematic representation of a synchronous demodulation scheme to detect carrier envelope containing magnetic field information

on synchronous demodulation, before analog to digital conversion, the envelope containing magnetic field information can be extracted (Fig. 9.2). Demodulation circuit depends on driving signal type: either a sine wave or a square wave. A typical demodulator for the first case is based on Gilbert cells with the main drawback of higher power consumption not negligible with respect to analog frontend. The latter, instead, can be demodulated using a set of switches to multiply by +1 and −1; it can be implemented using MOSFET gated by a reference signal controlled from oscillator. A sinusoid driving signal has main advantages to avoid noise folding and require a lower bandwidth in analog chain. Yet, circuitry to generate a sine wave and for related demodulation is usually more complex. Thinking of a final integrated system, the choice of a square wave based driving allows to save power consumption.

A final remark about system design is related to minimum and maximum magnetic field to be sensed and so the dynamic range of the system. A full scale range (FSR) of ±100 μT is requested according to given specifications (for magnetic field measurement with good linearity) but it is a good design strategy to increase further to cope with external magnetic field which may saturate the device:

$$\begin{aligned} \mathrm{FSR} &= \pm 1.2\ \mathrm{mT} \\ B_{\min} &\sim 1\ \mu\mathrm{T} \end{aligned} \tag{9.1}$$

and the corresponding dynamic range is:

$$\mathrm{DR} = 20 \cdot \log\left(\frac{2.4\ \mathrm{mT}}{1\ \mu\mathrm{T}}\right) = 67.6\,\mathrm{dB} \tag{9.2}$$

9.2 A Continuous-Time Fully Differential Transresistance Amplifier

An integrated circuit was implemented aiming at improving resolution and demonstrating the feasibility of a double-chip (MEMS + ASIC) magnetic field sensing system for consumer applications. A first prototype of VLSI, based on a topology similar to the one used for discrete components circuitry, has been designed in a standard CMOS 150 nm process. A negative feedback double-channel transresistance amplifier is designed using a fully differential operational amplifier; a simplified schematic is shown in Fig. 9.3. This choice rejects all mismatches that would occur when using two separate single-ended stages. In the differential configuration indeed offset drifts, temperature effects, and long term stability of the active component affect the two channels in the same way. In order to design the first analog stage to couple to MEMS the following blocks are required:

- high-gain, ultra-low-noise, fully differential amplifier with a suitable feedback network;
- common mode feedback network to set and stabilize both input and output biasing levels;
- output buffers to monitor signals without loading the outputs of the amplifier.

9.2.1 ASIC Noise

As calculated in Sect. 4.3, the overall noise budget for ASIC is:

$$\sqrt{S_{B_{\text{eln}}}} = \sqrt{S_{B_{\text{mech}}}} = \sqrt{\frac{S_{B\text{TOT}}}{2}} = 223\ \text{nT}/\sqrt{\text{Hz}}\,. \tag{9.3}$$

which must account for:

- Analog interface noise: the complete analog signal path between sensor and ADC (if it is a digital product) or output PAD (for an analog product);
- ADC quantization and thermal noise (if a digital output is needed);
- Digital post-processing noise (i.e., quantization noise).

ADC and digital post-processing will not be discussed in this manuscript but their noise must be accounted for. A possible design approach is to equally split noise budget between analog interface and ADC, under the assumption that digital noise is almost negligible. Alternatively, sensor analog interface sets the SNR of the full signal chain and ADC noise is negligible. This type of optimization depends on architectures, area, and power consumption considerations for each specific case. ASIC noise budget reported in Eq. (9.3) can be converted in terms of capacitance resolution according to:

$$\begin{aligned}\sqrt{S_{C,\text{eln}}} &= 2C_0 \cdot \frac{\sqrt{S_{x,\text{eln}}}}{g} \\ &= 2C_0 \cdot \frac{Q}{k} \cdot \frac{I \cdot L \cdot \sqrt{S_{B,\text{eln}}}}{g} \\ &= 158\ \text{zF}/\sqrt{\text{Hz}}\,,\end{aligned} \tag{9.4}$$

for an optimized device.

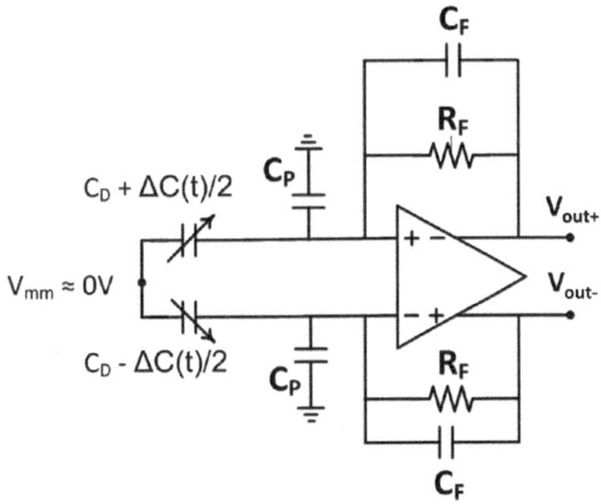

Fig. 9.3 Fully differential transresistance amplifier with MEMS modelled as a differential variable capacitor

9.2.1.1 Analog Interface Noise

The first amplifier in the signal path is typically the most important block to set the desired SNR: capacitive coupling to the sensor and the noise of the amplifier itself must be taken into account. It is now assumed that the full ASIC noise budget is accounted for the input stage, a fully differential transresistance amplifier reported in Fig. 9.3. This circuit is negative looped and so the two input pins of the operational amplifier can be considered virtual grounds whose value is fixed by a common mode feedback network (required in all fully differential schemes). Differential output voltage, $V_{\text{out}} = V_{\text{out}+} - V_{\text{out}-}$ using Laplace domain is:

$$
\begin{aligned}
V_{\text{out}+} - V_{\text{out}-} &= V_{\text{BIAS}} \cdot \left[\frac{1}{\frac{1}{s\left(C_0 - \frac{\Delta C(t)}{2}\right)}} - \frac{1}{\frac{1}{s\left(C_0 + \frac{\Delta C(t)}{2}\right)}} \right] \cdot \frac{R_F}{1 + sR_F \cdot C_F} \\
&= \frac{R_F}{1 + sR_F \cdot C_F} \cdot V_{\text{BIAS}} \left[sC_0 - s\frac{\Delta C(t)}{2} - sC_0 - s\frac{\Delta C(t)}{2} \right],
\end{aligned}
\tag{9.5}
$$

and with some reductions output voltage vs. input differential capacitance variation transfer function is obtained:

$$
\frac{V_{\text{out}}(t)}{\Delta C(t)} = -V_{\text{BIAS}} \cdot \frac{sR_F}{1 + sR_F \cdot C_F}. \tag{9.6}
$$

Ideally, output voltage does not depend on input parasitic capacitances and on MEMS rest value. This transfer function is useful to convert output noise into an equivalent input noise in terms of minimum differential capacitance variation which can be detected. The main noise sources in a fully differential continuous-time transresistance amplifier are given by feedback resistors and MOSFETs of input differential pair and related active load. These two contributions are now detailed.

Feedback Resistor Noise

Output voltage noise due to feedback resistors of differential transresistance amplifier is:

$$S_{\text{out},R_F,n} = 2 \cdot \frac{4k_B \cdot T}{R_F} \cdot \frac{R_F^2}{(1 + sR_F \cdot C_F)^2} \left[\frac{\text{V}^2}{\text{Hz}}\right], \tag{9.7}$$

where the factor 2 takes into account two independent noise sources from feedback resistors. Equation (9.7) can be converted into a corresponding $\Delta C_{R_F,n}$ using the input vs. output transfer function (Eq. (9.6)):

$$S_{C_R,n} = \frac{S_{\text{out},R_F,n}}{|T(j\omega)|^2} = \frac{8 \cdot k_B T}{R_F} \cdot \frac{1}{\omega_r^2 \cdot V_{\text{BIAS}}^2} \left[\frac{\text{F}^2}{\text{Hz}}\right], \tag{9.8}$$

Under the hypothesis that noise due input pair transistors of the operational amplifier is negligible with respect to noise of feedback resistors (this hypothesis is verified in Sect. 9.2.1.1), at the nominal frequency of 28.3 kHz the following values of R_F are estimated to meet noise analog interface noise specifications:

- $R_F \approx 4.6\ \text{M}\Omega$ considering a $V_{\text{BIAS}} = 3$ V;
- $R_F \approx 18.6\ \text{M}\Omega$ considering a $V_{\text{BIAS}} = 1.5$ V;

Here comes one of the main drawbacks of choosing a transresistance amplifier in an integrated CMOS technology: the implementation of high value resistors requires large active areas and possible solutions for this implementation can be done with:

- Integrated resistors;
- MOSFET working in ohmic region or subthreshold/off condition;
- Resistors implemented using switching capacitors.

A solution based on integrated resistors, compared to the implementation of a high ohmic resistance with MOSFETs in subthreshold/off region reg has main advantages in terms of better linearity, signal full scale, and noise with the drawback of occupied active area. Considering a standard CMOS technology with the possibility to implement polysilicon resistors with a sheet resistance in the range of 300 Ω/sheet, a rough estimation of the required area for a resistor value of about 5 MΩ is:

$$\left(\frac{L}{W}\right)_{\text{Poly-Si}} = \frac{2 \cdot R_F}{R_{\text{sheet}}} = \frac{2 \cdot 5 \cdot 10^6}{300} = 3.3 \cdot 10^5 \tag{9.9}$$

and so with a width dimension of 1 μm, the area occupied by feedback resistors is estimated to be approximately $A_{\text{Poly-Si}} = 190\ \mu\text{m} \times 190\ \mu\text{m}$. Width dimension must be chosen larger than minimum allowed dimension in order to have a good matching.

Input Stage Noise

In a typical (fully differential) operational amplifier main sources of noise are transistors of input pair and of its active load. These sources must now be compared to feedback resistor noise and, possibly, make them negligible. Voltage noise due to transistors is:

$$S_{V,\text{MOSFETs}} = \frac{2 \cdot 4\gamma \cdot k_B T}{g_m} \quad \left[\text{V}^2/\sqrt{\text{Hz}} \right] . \tag{9.10}$$

This voltage noise spectral density can be transferred to the output of the TIA and then converted into an equivalent resolution in terms of capacitance variation as done for R_F noise:

$$\begin{aligned} S_{C,\text{MOS}_n} &= \frac{2 \cdot 4k_B T \cdot \gamma}{g_m} \cdot \frac{1 + \omega^2 \cdot R_F^2 \cdot (C_F + C_{\text{in}})^2}{\omega_r^2 \cdot R_F^2 \cdot V_{\text{BIAS}}^2} \\ &= \frac{2 \cdot 4k_B T \cdot \gamma}{\sqrt{2 \cdot \mu C_{\text{ox}} \cdot \frac{W}{L} \cdot I_D}} \cdot \frac{1 + \omega^2 \cdot R_F^2 \cdot (C_F + C_{\text{in}})^2}{\omega_r^2 \cdot R_F^2 \cdot V_{\text{BIAS}}^2} . \end{aligned} \tag{9.11}$$

Considering a very conservative case due to the following strict working conditions:

- Width $W = 1.5\ \mu\text{m}$;
- Length $L = 0.9\ \mu\text{m}$;
- $\mu C_{\text{ox}} = 42\ \mu\text{A/V}^2$;
- $\gamma = 2$;
- Drain current $I_D = 0.5\ \mu\text{A}$;
- Resonance frequency $f_r = 28.3$ kHz;
- Feedback resistance $R_F = 5\ \text{M}\Omega$;
- Feedback capacitance $C_F = 90$ fF;
- Input parasitic capacitance $C_{\text{in}} = 2$ pF;
- Stators biasing voltage $V_{\text{BIAS}} = 3$ V;

an equivalent noise of about 76 $\text{zF}/\sqrt{\text{Hz}}$ is estimated which is, as previously supposed, negligible with respect to resistors noise. Noise due to the two transistor forming the active load of the input pair can be made negligible with an optimization of overdrive voltages and dimensions.

9.2.2 Fully Differential Operational Amplifier

Figure 9.4 shows the transistor level of a typical fully differential operational transconductance amplifier (OTA). The core transistors of the used technology are powered at 1.8 V which does not provide high enough swing to design a telescopic cascoded amplifier (this latter would have the main advantage of a lower power consumption). For this reason and to obtain a high open-loop gain a folded cascode amplifier is implemented.

In this configuration the open-loop gain of the amplifier is:

$$A_0 = \frac{V_{\text{out},a} - V_{\text{out},b}}{V_{\text{in},a} - V_{\text{in},b}} = g_{m_1} \cdot R_{\text{out}} \,, \tag{9.12}$$

and its bandwidth is:

$$f_p = \frac{1}{2\pi \cdot R_{\text{out}} \cdot C_{\text{out}}} \,, \tag{9.13}$$

being:

$$R_{\text{out}} = [g_{m_7} \cdot r_{0_7} \cdot (r_{0_3} \parallel r_{0_1})] \parallel [g_{m_9} \cdot r_{0_9} \cdot r_{0_1 1}] \tag{9.14}$$

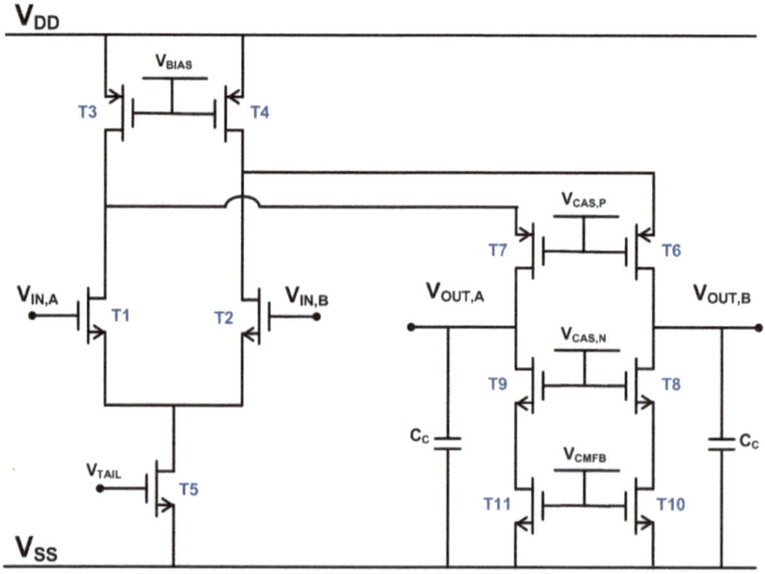

Fig. 9.4 Transistor level schematic of the fully differential operational amplifier

Below, some of the adopted design considerations to maximize the loop gain (90 dB) and minimize noise are summarized:

- a general constraint used in the design was to keep 15–20 mV of overdrive for all the transistors, to guarantee operation at least in the weak inversion region [2];
- the length of N-type input transistors pair T1 and T2 is chosen to avoid short channel effect ($\geq$2 μm). The N-type choice is due to their higher transconductance with respect to the P-type, while holding similar $1/f$ noise performances. Moreover, being the input signal modulated at >20 kHz
- the tail current is chosen to minimize the input-referred noise and to increase the overdrive voltage of transistors T3 and T4, so that their noise is almost negligible compared to the input pair. The same is done for the noise of T10 and T11, which are the cascode current generators. Input transistor areas are further increased to reduce the flicker noise;
- the biasing voltages of the cascode transistors (VCAS,N and VCAS,P in Fig. 9.4) are chosen to symmetrically increase the output dynamic range. The current in the cascode stage and the transistors sizes are set to achieve the desired gain as described above;
- finally, an estimation of a compensation capacitance at the outputs is done to guarantee a minimum phase margin of 60° in case of a buffer feedback. This compensation capacitance value will be revised considering common mode stability too.

Figure 9.5 shows the amplitude and phase of the open-loop gain with a compensation capacitance of 400 fF. The amplifier achieves a low, frequency open-loop gain of 90 dB, a GBWP of 4.4 MHz, a phase margin of 62°, and a current consumption of 4.4 μA.

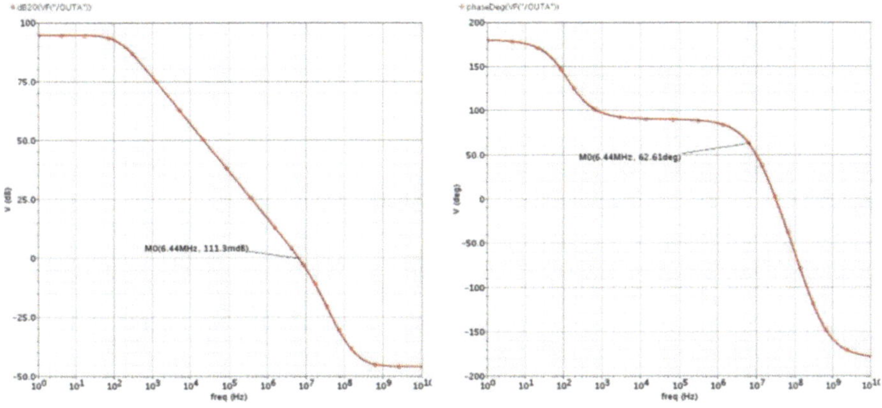

Fig. 9.5 Bode diagrams of fully differential folded cascode, with 400 fF compensation capacitance

9.2.2.1 Common Mode Feedback Network

As previously discussed, the magnetic field produces a differential signal between the input nodes; in order to reject all possible common mode contributions, that can lead to a shift in the polarization values—and can consequently change the behavior of the OTA—a common mode feedback network (CMFB) is needed.

Different solutions for the CMFB network can be implemented. The chosen solution monitors the output nodes, compares them with a reference voltage and acts on the folded cascode current generator to compensate the output DC level drift. The input nodes are DC stabilized by the feedback network. A schematic of this CMFB network is shown in Fig. 9.6.

If there is a positive common mode signal on the outputs of the amplifier, both T20 and T23 gatesource voltages are reduced, so that the currents in T27 and T26 are unbalanced, with $I_{T,27} < I_{T,26}$. The current of T26 is mirrored in the cascode current generators, reducing the current that flows on the outputs resistances. This leads in turn to a reduction of the amplifier outputs voltages, closing the negative feedback and correcting the common mode behavior of the OTA. In case of a differential signal, the sum of the currents that flows in T20 and T23 remains constant, and the same occurs in T26, so that the CMFB network circuit does not influence the differential behavior of the OTA. All the DC currents are chosen in order to have a 1:1 ratio in the folded cascode current mirror.

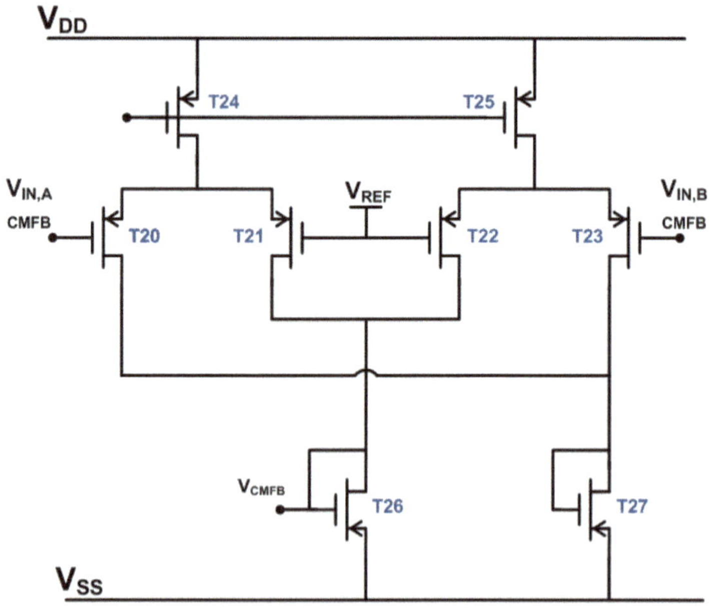

Fig. 9.6 Transistor level schematic of common mode feedback (CMFB) network. $V_{\text{in},a_{\text{CMFB}}}$ and $V_{\text{in},b_{\text{CMFB}}}$ correspond, respectively, to the outputs $V_{\text{out},a}$ and $V_{\text{out},b}$ of the differential amplifier of Fig. 9.4

The CMFB compensation capacitance is connected at the outputs of the fully differential amplifier, since this is the node with the highest impedance. It is possible to separate the compensation of the amplifier and of the CMFB network by connecting multiple capacitances either differentially between the two outputs or between the outputs and V_{SS}. The compensation capacitances are chosen to guarantee the stability of both differential and common mode amplifiers even in case of unity gain connection. An overall current consumption of less than 1 μA is estimated for this CMFB solution.

9.3 Verilog—A Model for Cadence® Environment

Models to simulate magnetometers (and more generally MEMS) taking into account physics, geometry, and technological constraints are implemented in a first phase of this project using Simulink® tool and they turned out to be very helpful in system design. It is however not straightforward to include in Simulink simulations secondary effects such as the presence of noise, parasitic capacitances, electronics nonidealities, etc.

In order to better analyze the overall performance and power dissipation from a fully coupled-system point of view and to prepare a custom tool for ASIC design, alternative solutions for sensor modelling are analyzed with an eye on languages compatible with integrated circuits simulators. Electromechanical coupled simulations are getting of deeper interest right for the design of ASIC for MEMS: sensor modelling inside one single simulation environment (Cadence®) allows to optimize electronic frontend considering mechanical nonidealities and sensor vs. electronics interference. As an ASIC must be designed inevitably using the software for which technology design-kit is available, a detailed magnetometer model has been integrated directly in the VLSI simulation environment. Verilog-AMS is a derivative of Verilog hardware description language which includes analog and mixed signal extensions to describe behavior of devices. For this specific case, Verilog-A, a continuous-time subset of Verilog-AMS, is used. A Verilog project consists of a hierarchy of modules, each one with a set of inputs, outputs, and bidirectional ports called "disciplines" (real, electrical, kinematic, thermal, magnetic, etc.). Each module can be associated to a schematic symbol the parameters of which can be changed in the object properties windows. A typical Verilog-A listing consists of:

- Inclusion of files required for code compilation;
- Additional "nature" declaration or change of default settings;
- Definition of modules inputs, outputs, and parameters;
- *analog begin/end* block where are reported sequential instructions which describe the model.

Two main blocks are developed to describe MEMS magnetometer behavior and their corresponding two symbols are placed in a schematic as if they were embedded libraries circuit parts, as shown in Fig. 9.7:

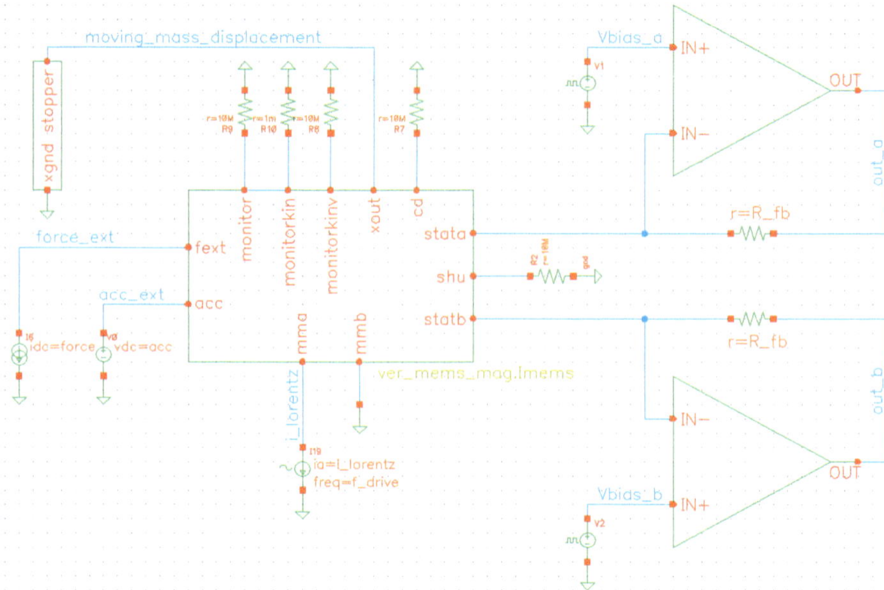

Fig. 9.7 Cadence® schematic where symbols (one for the sensors itself and one to simulate the mechanical stopper) associated to MEMS magnetometer are placed together with other electrical components

- *MEMSMAIN* which includes mechanical and electrical behavior of magnetometer, based on the balance of forces acting on the moving mass;
- *STOPPER* which is a behavioral implementation to simulate mechanical stoppers which are implemented to avoid moving mass collapsing to stators in case of pull-in or high acceleration shocks.

9.3.1 MEMS MAIN Module

The first implemented block is *MEMS MAIN*. Bidirectional ports which define this module are:

- *mma* and *mmb* are springs ends to which moving shuttle is anchored. They represent both inputs and outputs because Lorentz current is pumped into them but it is also interesting to monitor the resulting voltage because of their resistance. They are *electrical* discipline and so they are associated to Voltage "V" and Current "I";
- *stata*, *statb* are the two stators and *shu* is the rotor even though it does not actually have a direct connection with the package. This port is implemented to monitor the voltage of central point of flexures and so analyze effects due to springs mismatches. The two ports associated to readout electrodes are bidirectional as current due to motion must be sensed and a voltage must be forced. Also they are electrical.

- *fext* is an additional port to apply an external force different from Lorentz force. It is a *kinematic* discipline whose access functions are *Pos* (position) and *F* (force);
- *acc* is an input which allows to apply an external acceleration of value $g = 9.8\ \text{m/s}^2$. It is a *customkinematic* discipline which is not native in the language but it is implemented for the purpose of this model. Associated access function is*Accel*;
- *xout* represents moving mass displacement with respect to rest position. It is a *kinematic* output;
- *cd* is differential capacitance variation. It is a non-native *Electronics* discipline. The associated nature is *Cap*;
- *monitor*, *monitorkin*, and *monitorkinv* are auxiliary output for variables monitoring during design phase. They are, respectively, associated to: *electrical*, *kinematic*, and *kinematicv*(speed).

Parameters of this block which can be modified using the graphic interface of Cadence® environment (i.e., opening the object properties editor directly from the schematic) are:

- C_0, sensor rest capacitance;
- *gap*, gap between rotor and stators at rest;
- *damp*, damping coefficient;
- *mass*, effective mass of the device;
- *km*, elastic stiffness;
- *Lspring*, spring length;
- *B*, external magnetic field;
- *R1L*, *R2L*, *R1R*, and *R2R* resistances associated to each of four suspending elements.

Each of these parameters has a default value which can be modified according to the specific device and simulation; a range of allowed values is set (Fig. 9.8).

After this preliminary phase to define variables and parameters of the model, MEMS behavioral description is implemented. After calculating electrostatic forces both due to driving signal and stators biasing, balance of forces is solved and the resulting value of *xout* is used for capacitance variation calculation and current flowing through stators.

9.3.2 STOPPER Module

Stopper module is defined as a block with two bidirectional ports *xgnd* and *stopper* of *kinematics* type. This block monitors the displacement of the moving mass and in case of an external force generating pull-in, stopper block activates and limits the displacement to the value *diststop* which can be set in object properties and it is set, as default, to the same distance as implemented devices.

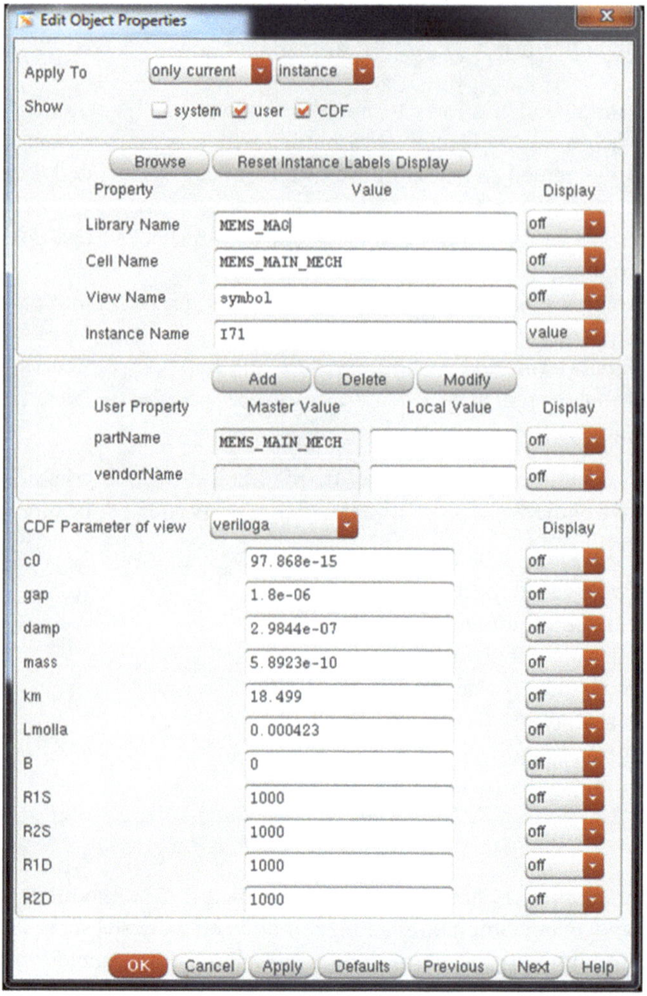

Fig. 9.8 The MEMS device can be added in the Cadence®—Virtuoso simulation tool as a basic building block of the circuit and its mechanical parameters can be set in object properties window

An *if/else* structure is used to distinguish when the moving mass is moving, in one direction or the other, beyond the maximum allowed displacement and, if that is the case, its movement is stopped. In order to check that behavioral implementation of the stopper is correct a sinusoid force is applied to the device to generate displacements larger than *diststop*. Figure 9.9 shows *xout* with respect to simulation time and it can be seen that when the moving mass reaches $diststop = 1.6$ μm, the displacement is stopped and limited as long as the force gets lower and the balance of forces gives a displacement lower than *diststop*.

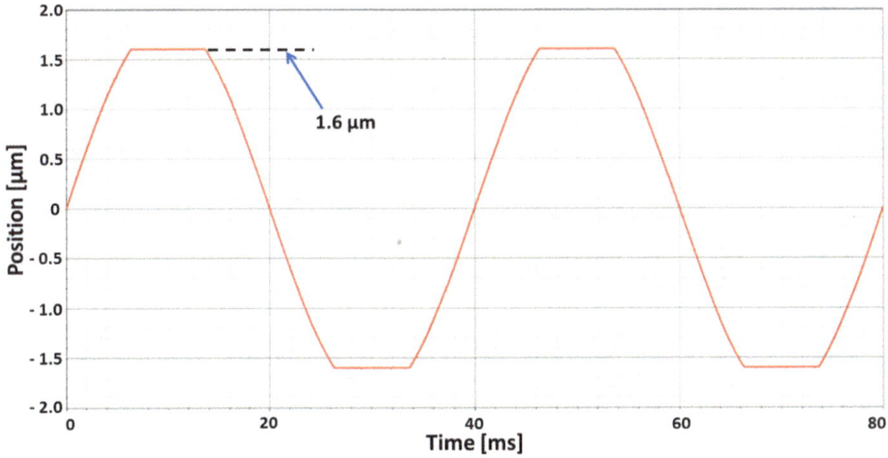

Fig. 9.9 Moving mass position with respect to time: when the maximum allowed displacement value is reached stopper module blocks the moving mass

9.3.3 Simulations and Model Verification

In order to verify a good implementation for the described Verilog-A model, a set of simulations are performed and more significant results are here reported. Simulations are based on magnetometer $N = 4$. As a first check, MEMS transfer function is simulated and reported in Fig. 9.10 with a correct value of resonance frequency, low-frequency value, and quality factor.

Further tests are executed in time domain comparing results with ones obtained by the same kind of simulation performed using Simulink®. For the sake of verification it is useful to compare Verilog-A model with the previously presented Simulink® model. Figure 9.11 reports two simulations, one performed using Simulink model and one with Verilog-A model in Cadence® environment, with an input acceleration. Indeed, this model is useful not only to simulate the device in presence of a magnetic field (purpose of this kind of sensor) but also in presence of other unwanted external forces, like accelerations.

9.4 Magnetic Field Measures: MEMS + ASIC

Magnetic field experimental measurements are performed with the implemented circuits and first generation of devices. MEMS is glued on a ceramic carrier and the ASIC is on top of the MEMS with stators wire-bonded to ASIC inputs and moving mass electrodes wire-bonded to carrier for external current excitation.

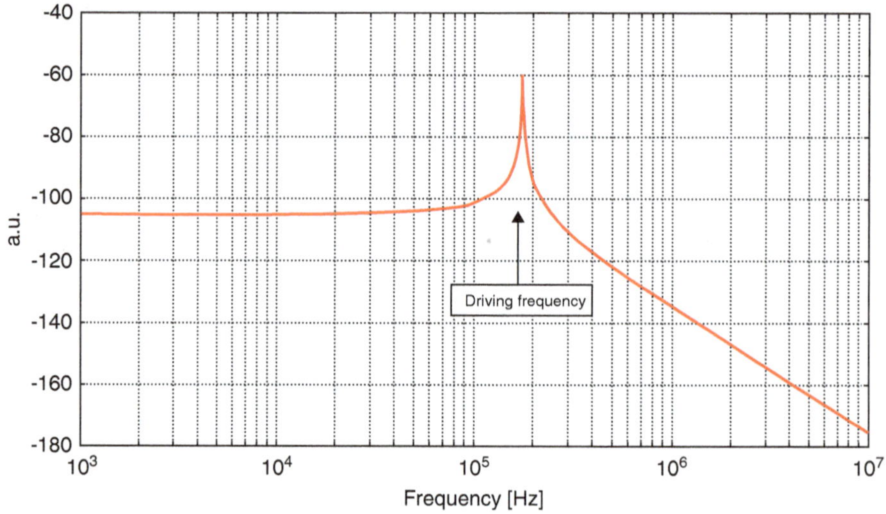

Fig. 9.10 Cadence® simulation using the Verilog-A magnetometer model, coupled to a CMOS frontend. The transfer function, solved by *Spectre* circuit simulator, is here simulated as the modulus of the differential capacitance variation vs. frequency

Figure 9.12 shows the transfer function of the fully differential transresistance amplifier for four different samples: resonance peaks are right below 30 kHz in agreement with measures previously done on devices and discrete electronics. Moreover, the circuit closed-loop bandwidth is at about 200 kHz in line with the pole set by feedback network.

Sensitivity measurements are performed applying three different values of magnetic field and measuring the resulting output signal of transresistance amplifier. Figure 9.13 shows, for three samples with 5 MΩ feedback resistance, that the peak is proportional to magnetic field intensity and measures have a good repeatability. As a comparison, one sample with 20 MΩ feedback resistor is reported and its output signal is four times higher than the others, on a fixed value of magnetic field. Measured sensitivity with 5 MΩ feedback resistors is 100 nV/μT in agreement with theoretical predictions and system simulations. ASIC resolution is set by feedback resistors noise and for a biasing voltage of 5 V corresponds to 200 $\mathrm{nT} \cdot \mathrm{mA}/\sqrt{\mathrm{Hz}}$.

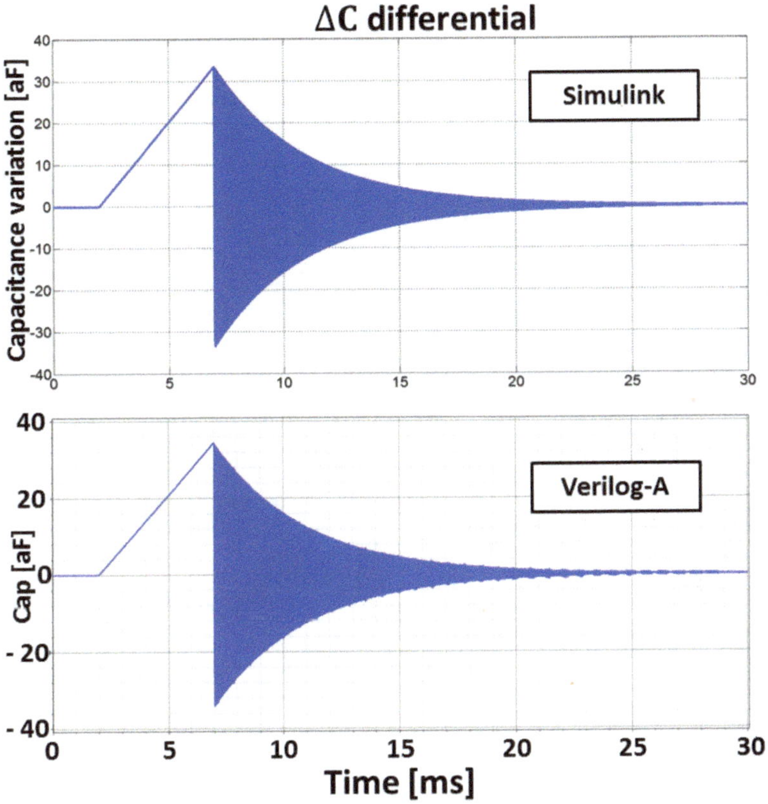

Fig. 9.11 Simulink and Verilog-A simulations of the effect of an external acceleration onto a magnetometer

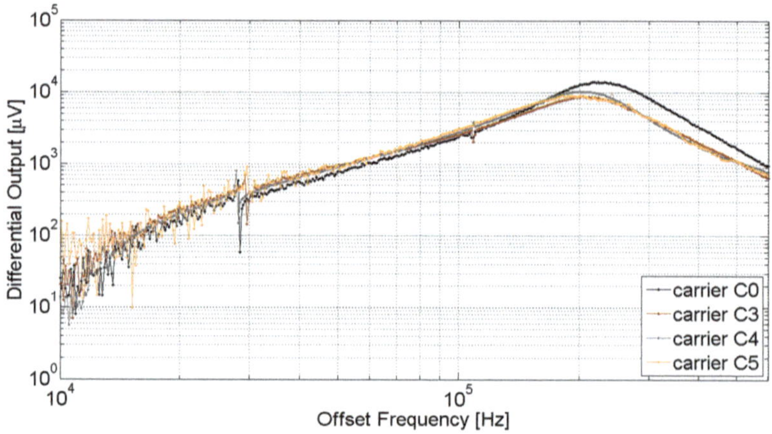

Fig. 9.12 Transresistance differential output voltage with respect to frequency for a fixed magnetic field. MEMS resonance peaks are close to 30 kHz and closed-loop bandwidth is at about 200 kHz in very good agreement with simulations

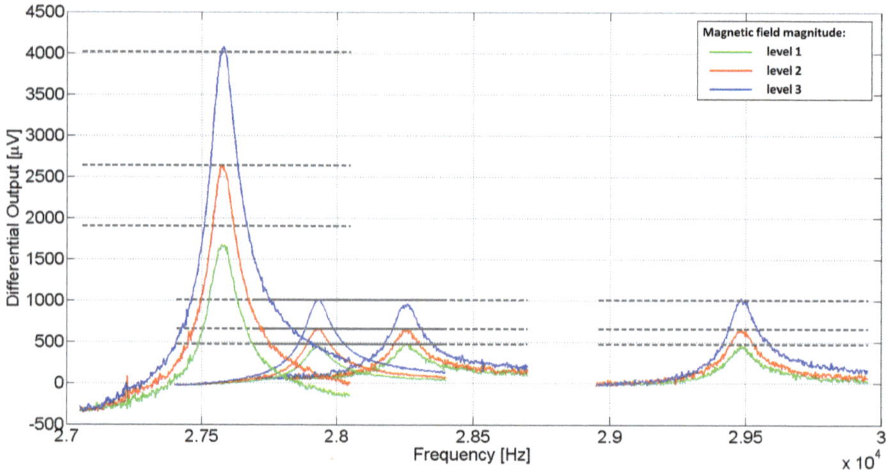

Fig. 9.13 Transresistance differential output voltage with respect to frequency. MEMS resonance peaks are close to 30 kHz and closed-loop bandwidth is at about 200 kHz in good agreement with simulations

References

1. B. Borovic, Q. Liu, D. Popa, H. Cai, F. Lewis, J. Micromech. Microeng. **15**(10), 1917 (2005)
2. P.R. Gray, P.J. Hurst, S.H. Lewis, R.G. Meyer, *Analysis and Design of Analog Integrated Circuits*, 5th edn. (Wiley, New York, 2009)

Chapter 10
From Prototype to Product

During the course of this book the full development of a MEMS magnetic field sensor prototype has been shown, providing some guidelines and highlighting critical points and potential pitfalls, as well as how to forecast or mitigate them. The feasibility of such a system based on multi-chip approach has been experimentally proven and the main steps, which represent the core of the design activity for a new product development, have been reported. Beside these ones, in an industry-related environment there are other standard development steps, including, for example, pads design, programming interface, biasing blocks, etc., in order to guarantee typical products specifications, such as ESD requirements, RFI immunity, etc., and to have high yield in production. All major steps needed will be mentioned in this chapter but they are not discussed in detail; however, some references are provided.

10.1 Towards Industrialization of MEMS Chip

As far as MEMS die is concerned, three major items are discussed:

- devices redesign;
- combination of several sensors in the same package;
- possible alternative working regimes for mechanical devices.

10.1.1 Devices Redesign: The Second Generation of Magnetometers

A second generation of devices is implemented with the same micromachining process, taking into account both obtained performance and the developed hardware to model correlation. Device architecture for Z-axis sensing element is the same as

C. Buffa, *MEMS Lorentz Force Magnetometers*,
DOI 10.1007/978-3-319-59412-5_10

previous generation and only magnetometers with $N = 4$ and $N = 8$ are designed because of their better performance as far as area, cross sensitivity to acceleration (they have lighter masses), and noise are concerned. If the pressure is kept in the range where free molecular flow dominates (and this is the case for industrial standard packages), the presented theory about sensitivity still applies. The main improvements of second generation of devices are:

- Longer beam: $L_{2\text{nd gen}} = 1060\ \mu\text{m}$; increasing springs length has a direct impact on increasing the mechanical sensitivity.
- Lower resonance frequency right at the border of audio bandwidth: $f_{r,2\text{nd gen}} = 20\ \text{kHz}$;
- A lower package pressure (taking into account technological constraints), possibly on order of 0.5 mbar.

Referring to Eq. (3.42), the corresponding expected improvement in terms of sensitivity is given by:

$$\frac{\frac{\Delta C}{\Delta B}_{2\text{nd gen}}}{\frac{\Delta C}{\Delta B}_{1\text{st gen}}} = \frac{L_{2\text{nd gen}}}{L_{1\text{st gen}}} \cdot \frac{f_{r,1\text{st gen}}}{f_{r,2\text{nd gen}}} \cdot \frac{b_{\text{area},1\text{st gen}}}{b_{\text{area},2\text{nd gen}}}$$
$$= \frac{1060\ \mu\text{m}}{868\ \mu\text{m}} \cdot \frac{28.3\ \text{kHz}}{20\ \text{kHz}} \cdot \frac{5.8\ \text{kg}/(\text{s}\,\text{m}^2)}{2.3\ \text{kg}/(\text{s}\,\text{m}^2)} = 4.36\ , \tag{10.1}$$

which corresponds to a better resolution in terms of minimum detectable magnetic field on equal current. Furthermore, the pumped current can now be reduced to save intrinsic power consumption and, in that case, the exact value should be trimmed to get a trade-off between resolution and power consumption. Figure 10.1 shows layouts of the second generation of devices.

10.1.2 Combination of Magnetometer with Other MEMS Sensors

Thanks to the developed theory and design methodology of mechanical devices, as reported in Chap. 3, an optimized magnetometer must be long but does not need to be large because its sensitivity is independent from the number of capacitive sensing cells. With such a form factor this device can fit in the same die where other MEMS sensors are integrated. A proposed possible solution is to combine it with a three-axis gyroscope, as shown in Fig. 10.2.

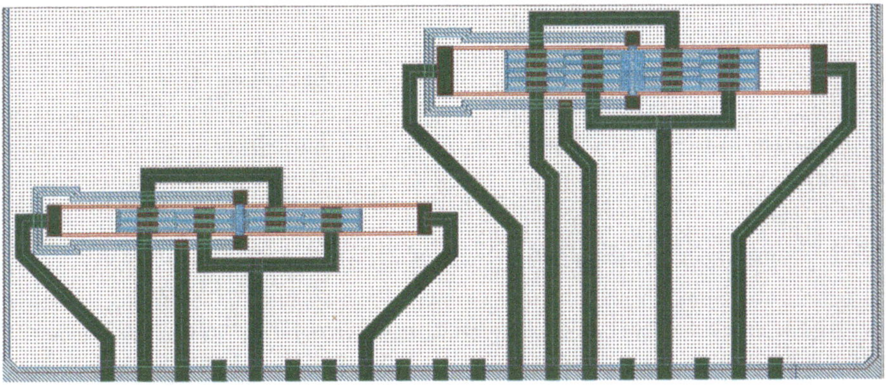

Fig. 10.1 Layout of the second generation of devices: beams length is extended to 1060 μm which is the maximum suspended dimension allowed to avoid buckling effects. Resonance frequency is fixed right at the border of audio bandwidth: 20 kHz. For $N = 4$ magnetometer, drawn spring width is $W_{s4,\text{2nd gen}} = 5.8$ μm and for $N = 8$ is $W_{s8,\text{2nd gen}} = 6.6$ μm

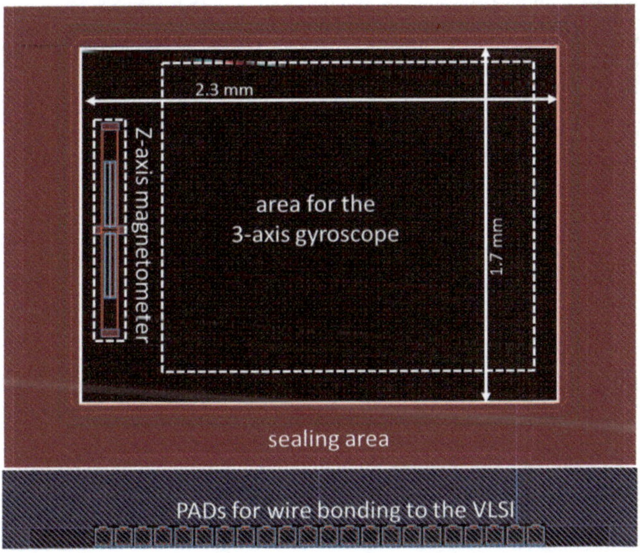

Fig. 10.2 Layout view of the Z-axis magnetometer inside a (2.3 mm × 1.7 mm) active area package. This high-aspect-ratio device can fit in a side of the same package of a three-axis gyroscope. A combination of this unit with a three-axis accelerometer results in a 7-DOF MEMS IMU that solves the problem of out-of-plane assembling of the Z-axis AMR magnetometer

10.1.3 Device Alternative Structures and Working Regimes

Device concept proposed in this manuscript is based on the excitation at resonance frequency of a main shuttle, to which capacitors are anchored, suspended by four springs (Fig. 3.7). At a later time, scientific research on MEMS magnetometer also investigated different architectures of devices and alternative operating modes; a comprehensive overview of these different implementations can be found in [1–3].

The choice among these different approaches may be done considering also product specifications on bandwidth and power consumption.

10.2 Towards Industrialization of ASIC Chip

In Chap. 9 ASIC design is presented, focusing on signal path. Towards the direction of a complete product, other blocks need to be implemented at ASIC level; a detailed discussion is not reported because they are blocks common to any sensor ASIC and they do not have a critical role in the development of a MEMS magnetometer. Among them, there are:

- Bandgap: providing with reference voltages for internal supplies, charge-pumps, and biasing of other blocks [4].
- Internal power supplies: typically generated with low dropout regulators (LDOs).
- ADC (as already discussed, in case of a digital output product). MEMS sensors applications typically require a maximum bandwidth of a few kHz and therefore sigma-delta modulators are extensively used [5].
- Design for testability: implementation of debugging features and the possibility to monitor internal nodes and references at one pad.
- Digital filtering and processing of signal.
- A digital part is also needed to control ASIC startup, calibration, and test features.
- ESD protections: particular attention is due to input pins where MEMS device is wire bonded to. Indeed these input nodes are critical for noise and parasitic capacitances. For ESD design, book [6] can be considered together with references which are there reported.

References

1. M. Li, S. Sonmezoglu, D.A. Horsley, J. Microelectromech. Syst. **24**(2), 333 (2015). doi:10.1109/JMEMS.2014.2330055
2. M. Li, D.A. Horsley, J. Microelectromech. Syst. **23**(6), 1477 (2014). doi:10.1109/JMEMS.2014.2316452
3. G. Langfelder, C. Buffa, P. Minotti, A. Longoni, A. Tocchio, S. Zerbini, in *2014 44th European Solid State Device Research Conference (ESSDERC)* (2014), pp. 62–65. doi:10.1109/ESSDERC.2014.6948758
4. P.R. Gray, P.J. Hurst, S.H. Lewis, R.G. Meyer, *Analysis and Design of Analog Integrated Circuits*, 5th edn. (Wiley, New York, 2009)
5. S. Pavan, R. Schreier, G.C. Temes, *Understanding Delta-sigma Data Converters*, 2nd edn. (Wiley-IEEE Press, New York, 2017)
6. V.A. Vashchenko, A. Shibkov, *ESD Design for Analog Circuits* (Springer, New York, 2010)

Chapter 11
Conclusions

This book deals with study and development of a compass for consumer applications. Solutions available on the market are based on technologies which require special magnetic materials with consequent higher costs for mass production. Scientific research about Z-axis element implemented using a standard industrial micromachining technology is of strong interest for the possibility to realize multi-axes IMUs in a single technological node.

In this manuscript a complete magnetic field sensing system based on multi-chip solutions (MEMS + ASIC) is presented. A flow and a methodology are also proposed to develop a prototype starting from some specifications towards a full product.

The main topics discussed in the book are:

- Development of a micromachined suspended architecture to detect magnetic field based on Lorentz force transduction principle.
- Development of a theory about mechanical sensitivity ($\Delta C/\Delta B$) and intrinsic noise (minimum detectable field due to thermomechanical noise) for devices working in free molecular flow regime, the typical working condition for devices requiring high Q in industrial packages. Under this hypothesis sensitivity turns out to be independent on the number of sensing cells N:

$$\frac{\Delta C}{\Delta B} = \frac{\varepsilon_0 \cdot I(t) \cdot L}{4\pi \cdot g^2 \cdot f_r \cdot b_{\text{area}}}, \tag{11.1}$$

and the minimum detectable magnetic field improves with a lighter device:

$$B_{\text{min}_{\text{int}}} = \frac{8 \cdot \text{BW}}{I \cdot L} \cdot \sqrt{k_{\text{B}} \cdot \pi \cdot T \cdot m}\,. \tag{11.2}$$

A new design approach is proposed: on a fixed resonance frequency, a tiny device performs better in terms of resolution and area without losing sensitivity.

C. Buffa, *MEMS Lorentz Force Magnetometers*,
DOI 10.1007/978-3-319-59412-5_11

- Development of a system level behavioral model using Simulink®. First generation of devices is designed according to the results of this model and supported by finite element simulation tools. Devices are fabricated using STMicroelectronics ThELMA micromachining process. An optimized device has an in-plane area occupation of 870 μm × 90 μm.
- Power budget partitioning and noise budget partitioning at a system level are discussed to set specifications for readout ASIC.
- Development of a custom instrument for precise electromechanical testing and devices characterization with a resolution of $\sim$1 aF/$\sqrt{\text{Hz}}$ and a bandwidth up to 100 kHz.
- Experimental validation of the proposed theory about sensitivity supported by static and dynamic electrostatic measurements. Measured damping coefficient of $b_{\text{area}} = 7.54\ \text{kg}/(\text{s}\,\text{m}^2)$ is $\approx$1.3 higher than expected. This is justified by a degassing of package pressure.
  System and damping models are validated through electromechanical characterization of devices.
- Development of an experimental setup for magnetic field measures: design of driving and readout electronics based on discrete components for device characterization in presence of a magnetic field. Driving circuitry is designed to minimize unwanted electrostatic forces.
  A measured sensitivity of 150 μV/μT, in line with the theoretical predictions of 195 μV/μT for a gap $g = 2.1$ μm is measured with designed readout electronics. Differences may result from a process under-etch with respect to the expected one. A resolution of 520 nT · mA/$\sqrt{\text{Hz}}$ is measured.
- Implementation of a Verilog-A model for electromechanical coupled simulations in Cadence® simulation environment.
- Magnetic field measurements with system prototype based on a multi-chip solution achieving a sensitivity of 100 nV/μT at the output of first amplifier with a feedback resistor of 5 MΩ. ASIC resolution is set by resistor noise.

Index

C. Buffa, *MEMS Lorentz Force Magnetometers*,
DOI 10.1007/978-3-319-59412-5

S

Kolibri 360 GmbH
Ferdinand-Jühlke-Straße 7
99095 Erfurt, Deutschland
produktsicherheit@kolibri360.de